THE LIFE OF ISAAC NEWTON

The Life of
Isaac Newton

Richard S. Westfall

CAMBRIDGE
UNIVERSITY PRESS

Published by the Press Syndicate of the University of Cambridge
The Pitt Building, Trumpington Street, Cambridge CB2 1RP
40 West 20th Street, New York, NY 10011-4211, USA
10 Stamford Road, Oakleigh, Victoria 3166, Australia

First published 1993
Canto edition 1994

Printed in the Unites States of America

Library of Congress Cataloging-in-Publication Data

Westfall, Richard S.
The life of Isaac Newton/ Richard S. Westfall.
p. cm.
Includes bibliographical references and index.
1. Newton, Isaac, Sir, 1642–1727. 2. Physics – History.
3. Science – History. 4. Physics – Great Britain – Biography
I. Title.
QC16.N7W34   1993
530'.092 – dc20
(B)   92-3377
                                                    CIP

A catalogue record for this book is available from the British Library.

ISBN 0-521-43252-9 hardback
ISBN 0-521-47737-9 paperback

TO

JOHN SAMUEL MCGRAIL
AND
BRIAN WESTFALL MCGRAIL

IN THE FOND HOPE THAT FIFTEEN OR TWENTY
YEARS FROM NOW, WHEN THEY ARE ABLE,
THEY WILL WANT TO READ THE BOOK

Contents

Preface

FEW MEN HAVE LIVED for whom less need exists to justify a biography. Isaac Newton was one of the greatest scientists of all times – and, in the opinion of many, not one of the greatest but the greatest. He marked the culmination of the Scientific Revolution of the sixteenth and seventeenth centuries, the intellectual transformation that brought modern science into being, and as the representative of that transformation he exerted more influence in shaping the world of the twentieth century, both for good and for ill, than any other single individual. We cannot begin to know too much about this man, and I will forbear to belabor the obvious and will say no more in justification of my book.

The life that I here present is a reduced version of the full-scale biography *Never at Rest*, which I published in 1980. In reducing the work in length, I have attempted to make it more accessible to a general audience by also reducing its technical content. (Very little mathematics appears in *The Life of Isaac Newton*. I invite those who feel the lack not only of mathematics but also of other technical details to consult the longer work.) To facilitate consultation, I have retained the titles of the original chapters; and the contents of the chapters, as condensations, follow the same patterns of organization. The numbers of the chapters do not correspond, however, for in condensing I have eliminated two of the fifteen in *Never at Rest* (Chapters 1 and 4). Chapter 4 dealt with Newton's development of his fluxional method or calculus; a summary of it appears in Chapter 3, "*Anni Mirabiles*," of the present book. It should then be easy to locate fuller discussions of any issue. The present *Life* also contains no footnotes. Anyone wanting to find the source of a particular quotation should be able quickly to locate it in *Never at Rest* in the same way.

Since publishing *Never at Rest,* I have moved on from Newton to other issues concerned with the history of early modern science and have not remained actively involved with Newtonian scholarship. Though I am aware of newer work that has appeared in the interim, I have not felt that I had rethought the issues sufficiently to attempt to incorporate it. Therefore *The Life of Isaac Newton* is not, as I have already indicated, a new work of scholarship but rather a shortened version of *Never at Rest.* I have included only one item that is not in the earlier work, Kenneth Baird's discoveries about Newton's maternal grandfather. (See K. A. Baird, "Some Influences upon the Young Isaac Newton," *Notes and Records of the Royal Society, 41* [1986–7], 169–79.) This extremely interesting information did not seem to me to require any refashioning of my presentation of Newton's childhood, and therefore I simply inserted it at the relevant location.

During the time when I was at work on Newton, I received assistance of many kinds from many sources. I acknowledged them once; I happily acknowledge them again. Grants from the National Science Foundation, the George A. and Eliza Gardner Howard Foundation, the American Council of Learned Societies, and the National Endowment for the Humanities, as well as sabbatical leaves from Indiana University, provided most of the time for study and writing, much of it in England, where the great bulk of Newton's papers exist. One of those years I had the privilege and advantage to be a Visiting Fellow of Clare Hall, Cambridge. The National Science Foundation and Indiana University also helped to finance the acquisition of photocopies of Newton's papers. The staffs of many libraries outdid themselves in kind assistance, most prominently (in proportion to my demands) the Cambridge University Library, the Trinity College Library, the Widener Library at Harvard, the Babson College Library, the Indiana University Library, and the Public Record Office. Most of the typing I owed to a succession of secretaries over the years in the Department of History and Philosophy of Science at Indiana University, but among them especially Karen Blaisdell. The help of Anita Guerrini in proofreading *Never at Rest* was invaluable, and the benefits of her assistance extend to the present volume. I cannot sufficiently express my gratitude to those I have mentioned and to many others who have helped in less central ways. I can at least try to express it, and I do.

Nor can any author omit his family. In 1980 I remarked that I had embarked on the biography of Newton by the time my children reached consciousness, and that I was finishing it as they completed their educations and set out on their own. The whole of their intimate experience of me was flavored by the additional presence of Newton. One of the happy changes wrought by the intervening years was the birth of the two grandsons to whom I dedicate this volume, while thanking all three children for their continuing encouragement and all of the joy they have brought to my life.

In the earlier work I also singled out my wife, as every male and married author surely must, and I underlined my gratitude by dedicating the book to her. She was completing a book of her own at the time. She is completing another as I publish this condensation of *Never at Rest*. There have been two others in between. I like to hope her scholarly production indicates that the support I have offered bears at least some small proportion to the support I have always enjoyed from her.

Acknowledgments

I WISH TO ACKNOWLEDGE permission granted me by the Trustees of the British Museum to reproduce pictures of the ivory bust by Le Marchand (Plate 5); by the University of California Press to reproduce a diagram from its edition of the English translation of the *Principia* (Figure 9); by the Bibliothèque Publique et Universitaire de Genève to reproduce the portrait of Nicolas Fatio de Duillier (Plate 2); by the Trustees of the National Portrait Gallery to reproduce the Kneller portrait of 1702 (Plate 3); by Neale Watson Academic Publications, Inc., to reproduce four diagrams from Richard S. Westfall, *Force in Newton's Physics* (London, 1971 (Figures 2, 3, 6, and 7); by Lord Portsmouth and the Trustees of the Portsmouth Estates to reproduce the Kneller portrait of 1689 (Plate 1) and the Thornhill portrait of 1710 (Plate 4); and by the Royal Society to reproduce the Vanderbank portrait of 1726 (Plate 6).

I wish further to acknowledge the permission and courtesy given me by Babson College (for the Grace K. Babson Collection); the Bodleian Library; the Syndics of the Cambridge University Library (for the Portsmouth Papers and other manuscripts); the University of Chicago Library (for the Joseph Halle Schaffner Collection); the Syndics of the Fitzwilliam Museum, Cambridge; the Jewish National and University Library (for the Yahuda manuscripts); the Provost and Fellows of King's College, Cambridge (for the Keynes manuscripts); the Warden and Fellows of New College, Oxford; the Royal Society; the Controller of H.M. Stationery Office (for crown-copyright records in the Public Record Office); and the Master and Fellows of Trinity College, Cambridge, to cite manuscripts.

The University of California Press has allowed me to quote from the Cajori edition of Newton's *Principia*; Cambridge University Press to quote from I. Bernard Cohen and Alexandre Koyré, eds., *Isaac Newton's Philosophiae Naturalis Principia Mathematica*; from B.J.T. Dobbs, *The Foundations of Newton's Alchemy*; from A. R. and M. B. Hall, eds., *Unpublished Scientific Papers of Isaac Newton*; from H. W. Turnbull et al., eds., *The Correspondence of Isaac Newton*; and from D. T. Whiteside, ed., *The Mathematical Papers of Isaac Newton*; Dover Publications, Inc., to quote from its edition of Newton's *Opticks*; Harvard University Press to quote from I. Bernard Cohen, ed., *Isaac Newton's Papers & Letters on Natural Philosophy*; Oxford University Press to quote from John Herivel, *The Background to Newton's 'Principia'*; and from Frank Manuel, *The Religion of Isaac Newton*; and *The Notes and Records of the Royal Society* to quote from R. S. Westfall, "Short-writing and the State of Newton's Conscience, 1662." I gratefully acknowledge all of their kindnesses.

A Note About Dates

BECAUSE ENGLAND had not yet adopted the Gregorian calendar (which it treated as a piece of popish superstition), it was ten days out of phase with the Continent before 1700, which England observed as a leap year, and eleven days out of phase after 28 February 1700. That is, 1 March in England was 11 March on the Continent before 1700 and 12 March beginning with 1700. I have not seen any advantage to this work in adopting the cumbersome notation 1/11 March, and the like. Everywhere I have given dates as they were to the people involved, that is, English dates for those in England and Continental dates for those on the Continent, without any attempt to reduce the ones to the others. In the small number of cases where confusion might arise, I have included in parentheses O.S. (Old Style) for the Julian calendar and N.S. (New Style) for the Gregorian.

In England the new year began legally on 25 March. Some people adhered faithfully to legal practice; many wrote double years (e.g., 1671/2) during the period from 1 January to 25 March. Everywhere, except in quotations, I have given the year as though the new year began on 1 January.

Plate 1. Newton at forty-six. Portrait by Sir Godfrey Kneller, 1689. (Courtesy of Lord Portsmouth and the Trustees of the Portsmouth Estates.)

Plate 2. Nicolas Fatio de Duillier. Artist unknown. (Courtesy of the Bibliothèque Publique et Universitaire de Genève.)

Plate 3. Newton at fifty-nine. Portrait by Sir Godfrey Kneller, 1702. (Courtesy of the Trustees of the National Portrait Gallery.)

Plate 4. Newton at sixty-seven. Portrait by Sir James Thornhill, 1710. (Courtesy of Lord Portsmouth and the Trustees of the Portsmouth Estates.)

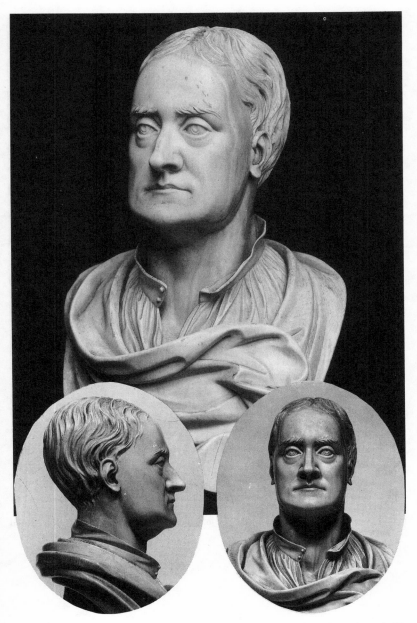

Plate 5. Newton at seventy-five. An ivory bust sculpted by David Le Marchand, 1718. (By permission of the Trustees of the British Museum.)

Plate 6. Newton at eighty-three. Portrait by John Vanderbank, 1726. (Courtesy of the Royal Society.)

A Sober, Silent,
Thinking Lad

ISAAC NEWTON was born early on Christmas Day 1642, in the manor house of Woolsthorpe near the village of Colsterworth, seven miles south of Grantham in Lincolnshire. Because Galileo, on whose discoveries much of Newton's own career in science would squarely rest, had died that year, a significance attaches itself to 1642. I am far from the first to note it – and undoubtedly will be far from the last. Born in 1564, Galileo had lived nearly to the age of eighty. Newton would live nearly to the age of eighty-five. Between them they virtually spanned the entire Scientific Revolution, the central core of which their combined work constituted. In fact, only England's stiff-necked Protestantism permitted the chronological liaison. Because it considered that popery had fatally contaminated the Gregorian calendar, England was ten days out of phase with the Continent, where it was 4 January 1643 the day Newton was born. We can sacrifice the symbol without losing anything of substance. It matters only that he was born and at such a time that he could utilize the work of Galileo and of other pioneers of modern science such as Kepler (who had been dead twelve years) and Descartes (who was still alive and active in the Netherlands).

Prior to Isaac, the Newton family was wholly without distinction and wholly without learning. As it knew steady economic advance during the century prior to Isaac's birth, we may assume that it was not without diligence and not without the intelligence that can make diligence fruitful. A Simon Newton, the first of the family to raise his head tentatively above rural anonymity, lived in Westby, a village about five miles southeast of Grantham, in 1524. Along with twenty-two other inhabitants of Westby, he had achieved the status of a taxpayer in the subsidy granted that year.

Fourteen of the twenty-two, including Simon Newton, paid the minimum assessment of 4d. Eight others paid assessments ranging from 12d to 9s 6d, and one, Thomas Ellis, who was one of the richest men in Lincolnshire, paid more than £16. If the Newtons had risen above complete anonymity, clearly they did not rank very high in the social order, even in the village of Westby. Because the average village in that part of Lincolnshire consisted of about twenty-five or thirty households, Simon Newton's assessment may indicate that he and thirteen others occupied the lowest rung on the Westby ladder. They were climbing, however, and rather rapidly. When another subsidy was granted in 1544, only four men from Westby had the privilege to pay; two of them were Newtons. Simon Newton was gone, but John Newton, presumably the son of Simon, and another John Newton, presumably his son, were now, after a man named Cony, the most flourishing inhabitants of Westby. In his will of 1562, the younger John Newton still styled himself "husbandman"; twenty-one years later, his son, a third John, died a "yeoman," a step up the social ladder; and a brother William of the same generation also claimed that standing.

Inevitably, Newton's pedigree has been worked out in considerable detail, first by Newton himself, later by the antiquarians whose attention the great attract. A list of his uncles, great-uncles, and the like and the relationships in which they stood to him are of less interest than the implications wrapped up in the shift from husbandman to yeoman. In Lincolnshire, the sixteenth and seventeenth centuries witnessed a steady concentration of land and wealth with a consequent deepening of social and economic distinctions. The Newtons were among the minority who prospered.

Westby is located on a limestone heath, the Kesteven plateau, a wedge of high ground thrust up toward Lincoln between the great fens to the east and the fenny bottom lands of the Trent valley to the west. The plateau had always presented itself as a likely highway to the north. The Romans had built Ermine Street along its back, and the Great North Road of medieval and early modern England followed the same route as far as Grantham, where it veered off to the west toward an easier passage over the Humber. Even today the main highway north near the eastern coast of England crosses the plateau along the same path. Woolsthorpe, where Newton was reared, lay less than a mile from a major thoroughfare of his day.

If the plateau was a natural highway, it was not a natural granary. The soil was thin and poor. Much of the arable land could sustain only a two-field rotation, which allowed it to stand fallow half the time. Enclosure here proceeded slowly, while large stretches of uncultivated waste were used in common as sheep walks. Wool from sheep was the foundation of the plateau's agricultural economy. In compensation for niggardly soil, the plateau bore a relatively sparse population. Those who would could prosper. The Newtons would.

The tale is told in the details of successive wills. From John Newton of Westby, who left a will when he died in 1562, each generation for a century left a considerably augmented estate. Rather, they left augmented estates. The Newtons were also a prolific clan. John Newton of Westby had eleven children, of whom ten survived. His son Richard, Isaac's great-grandfather, had seven children, of whom five survived. Isaac's grandfather, Robert, had eleven, of whom six survived. No single inheritance was augmented and passed on. The inheritance was continually being divided, but most of the segments took root and flourished. By the middle of the seventeenth century a considerable number of substantial yeomen named Newton were sprinkled over the area around Grantham, all of them descendants of John Newton of Westby, husbandman. No doubt the fact that this John Newton married very well – Mary Nixe, the daughter of a prosperous yeoman – helped his position. He must also have known how to handle the dowry, however, for he was able to provide handsomely for three sons. The descendants of one of them, William, prospered even more than the rest; in 1661 one of his descendants, yet another John, pushed his way into the squirearchy as Sir John Newton, Bart. In 1705, Isaac Newton anxiously pursued his son, also Sir John Newton, Bart., for corroboration of his pedigree. At about the time of his death, John Newton of Westby purchased an extensive farm of well over a hundred acres, including sixty acres of arable land, in Woolsthorpe for another son, Richard. Woolsthorpe lay approximately three miles south-west of Westby, and Richard Newton was Isaac Newton's great-grand-father. To put the family's economic position in perspective, the average estate in the 1590s of peasants on the heath with property – that is, of the wealthier peasants – was about £49. The richest yeoman, as measured by his will, who died in Lincolnshire in the 1590s left personal property of nearly £400. Very few wills in that time left goods worth more than £100.

Richard Newton, whose father established him on a farm purchased for £40, left goods inventoried at £104; the inventory did not include the land or the house. It did include a flock of fifty sheep, well over the average number. Sheep were the measure of wealth on the heath. Not only did John Newton of Westby endow three sons magnificently by yeoman standards, but he also married a daughter to Henry Askew (or Ayscough) of Harlaxton. The Ayscoughs were a prominent Lincolnshire family, though it is not clear what, if any, relation Henry Askew bore to the main stem of the family, whose seat lay well to the north. It was not the last alliance between the two families.

Robert Newton, Isaac's grandfather, was born about 1570. He inherited his father's property at Woolsthorpe, to which he added the manor of Woolsthorpe by purchase in 1623. The manor was not in prosperous condition. It had changed hands by sale four times during the previous century. Nevertheless its value was reckoned at £30 per year. Added to the original estate, it gave the family a comfortable living indeed by yeoman standards of the day. Socially, it may have elevated Robert still further. He was now lord of a manor, legally entitled to exercise the powers of local authority, such as conducting court baron and court leet, which, as still operative elements of local administration, had jurisdiction over minor breaches of the peace and could levy fines but could not imprison. The lord of a manor was no husbandman. In December 1639, he settled the entire Woolsthorpe property on his eldest surviving son, Isaac, and Hannah Ayscough (or Askew), to whom Isaac was betrothed. Isaac was hardly a young man. He had been born on 21 September 1606. Although Hannah Ayscough's age is not known, it seems likely that she was well beyond maidenhood herself; her parents had been married in 1609, and her brother William was probably the William Askue who matriculated in Cambridge from Trinity College early in 1630. Nevertheless, the couple did not marry at once, and there is every suggestion that they waited to obtain the inheritance first. After all, Robert Newton was nearly seventy. He obliged in the autumn of 1641; the following April the two were united.

The Ayscough match was another distinct step forward for the Newtons. Hannah was the daughter of James Ayscough, gentleman, of Market Overton, county Rutland. As marriage portion, she brought with her a property in Sewstern, Leicestershire, worth £50 per year. It is difficult to imagine the match without Newton's recently purchased dignity of manorial lord. Hannah brought more than additional wealth. For the first time, the Newtons made contact with formal learning. Before 1642, no Newton in Isaac's branch of the family had been able to sign his own name. Their wills, drawn up by scriveners or curates, bore only their marks. Isaac Newton, the father of our subject, was unable to sign his name, and so was his brother who helped prepare the inventory of his possessions. In contrast, one Ayscough at the very least was educated. William, Hannah's brother (M.A. Cambridge, 1637), pursued the calling for which learning was essential. Ordained to the clergy of the Anglican Church, he was instituted to the rectory of Burton Coggles, two miles east of Colsterworth, in January of the year in which his sister married Isaac Newton.

As it turned out, Isaac was reared entirely by the Ayscoughs. We can only speculate what would have happened had his father lived. The father was now the lord of a manor, as his own father had not been while he himself was being reared. Perhaps he would have seen the education of his son as a natural consequence of his position. However, his brother Richard, who was only a yeoman to be sure and not the lord of a manor, did not see fit to educate his son, who died illiterate. Being reared as an Ayscough, Isaac met a different set of expectations. The presence of the Reverend William Ayscough only two miles to the east may have been the critical factor. At a later time, his intervention helped to direct Isaac toward the university. Whatever the individual roles, the Ayscoughs took it for granted that the boy would receive at least a basic education. We have some reason to doubt that the Newtons would have done so.

Six months after his marriage, Isaac Newton died early in October 1642. He left behind an estate and a pregnant widow but virtually no information about himself. We have only one brief description of him, from a century and a half after his death, by Thomas Maude, who claimed to have inquired diligently into Newton's ancestry among the descendants of his half-brother and half-sisters and around the parish of Colsterworth. According to Maude, Isaac Newton the father was "a wild, extravagant, and weak man." Such may have been the case; because Maude did not even get his name right, however, calling him John, we are scarcely compelled to accept the description. About his estate we are directly informed by his will. Because it defines the economic position of Isaac

Newton (the son) at the time of his birth, it deserves some scrutiny. In addition to his extensive lands and the manor house, Isaac Newton, senior, left goods and chattels valued at £459 12s 4d. His flock of sheep numbered 234, which compares with an average flock of about 35. He owned apparently 46 head of cattle (divided among three categories which are partly illegible on the document and hard to interpret in any case), also several times the average. In his barns were malt, oats, corn (probably barley, the staple crop of the heath), and hay valued at nearly £140. Because the inventory was made in October, these items undoubtedly represent the harvest of 1642. By putting oats (£1 15s) in a separate category but coupling corn and hay (£130) in another, the men who drew up the inventory made it difficult to interpret. The oats and hay would have been fodder for the winter; surely the corn would not have been. The cattle (worth £101) and the sheep (worth £80) would have consumed the fodder during the approaching winter, so it does not constitute a final product of the estate. Part of the final product was wool, and the inventory includes wool valued at £15. It is unlikely that the 1642 clip, from June, would still have been on hand; £15 is too small a sum in any case, as the annual clip averaged between a fourth and a third of the value of the flock. The estate also included, of course, extensive agricultural equipment and furnishings for the house. It included as well rights to graze sheep on the common. The value of such rights is impossible to estimate, but when wool is king, grazing rights are gold. Like fodder, of course, they would only be means to the annual product, however. At this remove it is impossible to determine the total annual value of the estate. An estimate of at least £150 per year does not seem unreasonable. We should add that the inventory may have been lower than the long-term average value of the estate. The 1620s had been a hard decade, and probated inventories throughout the 1630s were lower in consequence. They did not fully recover their former level until about 1660. Newton's mother reserved the income from the paternal estate to Isaac when she remarried; the dowry lands in Sewstern appear to have been included. In addition, her second husband settled a further piece of land on him. Ultimately Newton inherited the entire paternal estate together with the land from his stepfather and some additional properties purchased by his mother. I have summarized the estate in financial terms because that was its only meaning in Newton's life. At one time the family intended that he manage it. This was not to be, however, and the estate functioned in his life only as financial security. Whatever problems might await the child still unborn when the inventory was made, poverty was not likely to be among them.

❧

The only child of Isaac Newton was born three months after his father's death in the manor house at Woolsthorpe early Christmas morning. The posthumous offspring, a son, was named after his father, Isaac. Already fatherless, apparently premature, the baby was so tiny that no one expected him to survive. More than eighty years later, Newton told John Conduitt, the husband of his niece, the family legend about his birth. Conduitt tells us:

Sr I. N. told me that he had been told that when he was born he was so little they could put him into a quart pot & so weakly that he was forced to have a bolster all round his neck to keep it on his shoulders & so little likely to live that when two women were sent to Lady Pakenham at North Witham for something for him they sate down on a stile by the way & said there was no occasion for making haste for they were sure the child would be dead before they could get back.

Apparently his life hung in the balance at least a week. He was not baptized until 1 January 1643.

We expect little information on the following years, and we are not disappointed. We do know, however, about one event of overwhelming importance which shattered the security of Newton's childhood immediately following his third birthday. Conduitt obtained an account of it from a Mrs. Hatton, née Ayscough:

Mr Smith a neighbouring Clergyman, who had a very good Estate, had lived a Batchelor till he was pretty old, & one of his parishioners adviseing him to marry He said he did not know where to meet with a good wife: the man answered, the widow Newton is an extraordinary good woman: but saith Mr Smith, how do I know she will have me. & I don't care to ask & be denyed. But if you will go & ask her, I will pay you for your day's work. He went accordingly. Her answer was, She would be advised by her Bro: Asycough. Upon which Mr Smith sent the same person to Mr Ayscough on ye same errand, who, upon consulting with his Sister, treated with Mr Smith: who gave her son Isaac a parcell of Land, being one of the terms insisted upon by the widow if she married him.

Barnabas Smith was the rector of North Witham, the next village south along the Witham, a mile and a half away. Born in 1582, he had matriculated at Oxford in 1597, commencing B.A. (as graduation was called at the time) in 1601 and proceeding M.A. in 1604. "Pretty old," as Mrs. Hatton's account has it, rather understates the matter; he was sixty-three years old when he added "Smith" to Hannah Ayscough Newton's lengthening string of names. Nor had he lived a bachelor. He had buried a wife the previous June, and he had not allowed much grass to grow over her grave before he mended his single estate.

We do not know a great deal about the rector of North Witham. To start with the best, he owned books. Newton's room at Woolsthorpe contained, on the shelves that Newton had built for them, two or three hundred books, mostly editions of the Fathers and theological treatises, which had belonged to his stepfather. Purchasing books with intent to study is, of course, not the only way to obtain them. One might inherit a theological library, for example, if one's father was a clergyman as Barnabas Smith's father was. At any rate he had the books. He may even have read a bit in them. In a huge notebook, which he began in 1612, Smith entered a grandly conceived set of theological headings and under the headings a few pertinent passages culled from his reading. If these notes represent the sum total of his lifetime assault on his library, it is not surprising that he left no reputation for learning. Such an expanse of blank paper was not to be discarded in the seventeenth century. Newton called it his "Waste Book," and what Barnabas Smith had once intended as a theological commonplace book witnessed the birth of the calculus and Newton's first steps in mechanics. Possibly the library started Newton's theological voyage to lands his stepfather would not have recognized.

Smith must have been vigorous, not to say lusty; though already sixty-three when he married Hannah Newton, he fathered three children before he died at seventy-one. No surviving story suggests that he concerned himself much with the likelihood that the three children would soon be left without a father, even as another boy had been. Beyond the books and the vigor, little else about him sounds attractive. He occupied the rectory in North Witham because his father, the rector of South Witham, had bought it for him in 1610 by purchasing the next presentation from Sir Henry Pakenham, who controlled it. In the following year, a visitation by the Bishop of Lincoln reported that the Reverend Mr. Smith was of good behavior, was nonresident, and was not hospitable. In effect, Barnabas Smith's father had purchased a comfortable annuity for his son.

He received the income from North Witham for more than forty years. For the first thirty, as far as we know, he conformed without protest to the ever more Arminian policies of the established church. With the Civil War came the Puritans and the Covenant. The Reverend Mr. Smith remained undisturbed in his living. The second Civil War brought the Independents and the Engagement. By now, large numbers of steadfast Anglican clergy had preferred ejection to conformity, and many were suffering real deprivation, but not the Reverend Mr. Smith. When he died in 1653, he had grasped his living firmly through all the upheavals – a pliable man, obviously, more concerned with the benefice than with principles. Although they had never met, John Milton knew him well.

> Anow of such as for their bellies sake,
> Creep and intrude, and climb into the fold?
> Of other care they little reck'ning make,
> Than how to scramble to the shearers feast,
> And shove away the worthy bidden guest;
> Blind mouthes!

Not that the living of North Witham was Barnabas Smith's primary means of support. He had an independent income of about £500 per annum – "wch in those days was a plentiful estate . . . ," said Conduitt in his single essay at understatement. For Newton, his stepfather's wealth meant in the end a significant increment to his own possessions. As Mrs. Hatton's account states, part of the marriage settlement was a parcel of land for him, increasing his paternal estate. Years later, Newton inherited from his mother additional lands that she had purchased for him, undoubtedly from the estate of her second husband. The will of Newton's uncle, Richard Newton, suggests an economic status similar to that of Newton's father. The will of Hannah Ayscough Newton Smith bespeaks a wholly different level. The Ayscough marriage had been a step upward for the Newtons, more in status than in wealth. The Smith marriage brought a major increase in wealth. In return, it deprived Newton of a mother. His stepfather had no intention of taking the three-year-old boy with the mother. Isaac was left in Woolsthorpe with his grandparents Ayscough. The Reverend Mr. Smith did have the house rebuilt for them; he could afford it.

The loss of his mother must have been a traumatic event in the life of the fatherless boy of three. There was a grandmother to replace her, to be sure; but significantly, Newton never recorded any affectionate recollection of her whatever. Even her death went unnoticed. Even more significant is the grandfather. Until very recently, everyone assumed that the grandmother was a widow because there is no single reference to him in Newton's papers. We now know that the grandfather was present in the manor house as well. We also know that he returned Isaac's affection in full measure – that is, he excluded him entirely from his will.

As we shall see, Newton was a tortured man, an extremely neurotic personality who teetered always, at least through middle age, on the verge of breakdown. No one has to stretch credulity excessively to believe that the second marriage and departure of his mother could have contributed enormously to the inner torment of the boy already perhaps bewildered by the realization that he, unlike others, had no father. Moreover, there is reason to think that Isaac Newton and Barnabas Smith never learned to love each other. Nine years after his stepfather's death in 1653, when Newton was moved to draw up a list of his sins, he included "Threatning my father and mother Smith to burne them and the house over them." Probably every boy has angry confrontations with his parents, when puerile threats are screamed in frustration. Nevertheless, the scene must have etched itself deeply on Newton's consciousness if he recalled it nine years later. For Barnabas Smith's part, his actions speak clearly enough. For more than seven-and-a-half years, until he died, while the child of three grew to be a boy of ten, he did not take him to live in the rectory in North Witham.

The manor house of Woolsthorpe stands on the west side of the small valley of the river Witham, a string down the Kesteven plateau beaded with villages, leading toward the town of Grantham. Built of the gray limestone that also builds the plateau, the house forms a squat letter T, with the kitchen in the stem, and the main hall and a parlor in the cross-stroke. The entrance, somewhat off-center between the hall and the parlor, faces the stairway, which leads upstairs to two bedrooms. Here Newton was born, and here was the room he occupied while he grew to adolescence. Beyond the fact that he attended day schools in the neighboring villages of Skillington and Stoke, we know little about his youth. The area was liberally sprinkled with aunts, uncles, and cousins of varying degrees. Wills tell us of two uncles Newton, one living in Colsterworth and one in Counthorpe three miles away, both with children apparently not far removed from Isaac in age. Three married aunts, all with children, lived in neighboring Skillington. There were also Dentons,Vincents, and Welbys who were more distantly related on the Newton side of the family. At least some connection with them was maintained; in the affidavit that accompanied and justified his pedigree in 1705, he stated that his grandmother Ayscough "frequently conversed with [his] great Uncle, Richard Newton" at Woolsthorpe. There were Ayscoughs as well. His grandmother had grown up in the area, and besides her daughter Hannah there was another married daughter, Sarah, not to mention the Reverend William Ayscough two miles away. Nevertheless Isaac's boyhood appears to have been lonely. He formed no bond with any of his numerous relations that can be traced in his later life. The lonely boyhood was the first chapter in a long career of isolation.

In August 1653, the Reverend Barnabas Smith died, and Newton's mother returned to Woolsthorpe to live. Perhaps the period that followed was a joyful interlude for the boy of ten, to whom a mother had been restored. Perhaps some bitterness tinged his joy as a half-brother and two half-sisters shared her attention, one an infant not yet a year old and another, just two, possibly dominating it. The fact is we do not know. We know only that the interlude was short. In less than two years' time, Isaac was sent off to grammar school in Grantham.

By Newton's own testimony, he entered the school in Grantham when he was twelve. The considerable number of anecdotes about this period of his life concentrate on his academic progress and on extracurricular recreations. By telling us nothing at all about the nature of his studies, they leave us to assume that he studied what every boy in grammar school at that time studied as a matter of course: Latin and more Latin, with a bit of Greek toward the end and no arithmetic or mathematics worth mentioning – such was the standard curriculum of the English grammar school of the day, and such we must assume Newton confronted in the school at Grantham, which was respected, and which Mr. Stokes, reputed to be a good schoolmaster, ran. The very silence on such a vital question by the collectors of anecdotes strongly suggests that Newton's education differed in no way from the ordinary one, and some of the earliest surviving fragments of Newtoniana confirm this. In 1659 he purchased a small pocketbook (or notebook, as we would say), dating his signature on the first page below a Latin couplet with "Martij 19, 1659." If one assumes this means 1659/60, it belongs to the period when he was back at Woolsthorpe. He devoted most of the notebook to "Utilissimum prosodiae supplementum." Further, in the Keynes Collection in King's College there is an edition of Pindar with Newton's signature and the date 1659, and the Babson Collection has his copy of Ovid's *Metamorphoses* dated that year.

The reader in the twentieth century, surrounded by the achievements of modern mathematics and the material culture it has generated, can scarcely believe that the man who would discover the calculus four years after he left grammar school was probably not even introduced there to the already thriving mathematical culture out of which the calculus would come. Neither is there any suggestion that he studied natural philosophy. Nevertheless, the grammar school in Grantham served Newton well. Without exception, the mathematical works on which he fed a few years hence were written in Latin, as were most of his sources in natural philosophy. Later still, he could communicate with European science because he wrote Latin as readily as English. A little arithmetic, which he could have absorbed in a day's time anyway, would scarcely have compensated for a deficiency in Latin.

One other important feature of the grammar school in the seventeenth century was the Bible. Studied in the classical tongues, it both supported the basic curriculum and reinforced the Protestant faith of England. In Newton's case, biblical study may have joined with the Reverend Smith's library to launch his voyage over strange theological waters.

In Grantham, Newton lodged with the apothecary Mr. Clark, whose house stood on the High Street next to the George Inn. Also living in the house were three stepchildren of Mr. Clark, named Storer from his wife's first husband: a girl whose first name has been lost, and two boys, Edward and Arthur. It seems clear that Newton did not get along with the boys. Among the incidents that he remembered uncomfortably in 1662 were "Stealing cherry cobs from Eduard Storer" and "Denying that I did so." He also recalled "Peevishness at Master Clarks for a piece of bread and butter." As far as we know, Newton had grown up in relative isolation with his grandparents. He was different from other boys, and it is not surprising if he was unable to get along with them easily. As they came to recognize his intellectual superiority, the boys in the school apparently hated him. Years later there was only one, Chrichloe, whom he remembered with pleasure. William Stukeley, a younger friend of Newton, who collected information about him while resident in Grantham in the eighteenth century, gathered that the others found him too cunning, able to get the better of them with his greater quickness of mind. Perhaps it was one such incident, hardly calculated to endear him to boys already hostile, that Newton recorded in 1662: "Putting a pin in John Keys hat on Thy day to pick him."

The stories that Stukeley collected in Grantham in the 1720s stressed the fact that Newton preferred the company of girls. For Miss Storer, who was several years his junior, and her friends he made doll furniture, delighting in his skill with tools. Indeed, as the two grew older, something of a romance apparently developed between Newton and Miss Storer. It was the first and last romantic connection with a woman in his life. The romance of an adolescent boy who prefers the company of girls is not likely to endure. This one did not. Though Newton remembered Mrs. Vincent (her married name) as one of his two friends in Grantham, it was only Mrs. Vincent who told of the romance. For the most part, he kept his own company. He was always "a sober, silent, thinking lad," Mrs. Vincent recalled, and "never was known scarce to play with the boys abroad."

Early in Newton's stay in Grantham, a crisis occurred which burned deeply into his memory. He had not even had time to assert his intellectual prowess. Whether because he was ill prepared by the village schools, or because he was alone again and frightened, he had been placed in the lowest form, and even there he stood next to the bottom. On the way to school one morning, the boy next above him kicked him in the belly, hard. It must have been Arthur Storer. Boys will be boys, but even among boys a vicious kick in the stomach requires some provocation. Already there may have been one too many peevish scenes over bread and butter and cherry cobs and all the rest one can imagine. Though he played with the girls, Newton knew what he had to do. In Conduitt's account:

[A]s soon as the school was over he challenged the boy to fight, & they went out together into the Church yard, the schoolmaster's son came to them whilst they were fighting & clapped one on the back & winked at the other to encourage them both. Tho Sr Isaac was not so lusty as his antagonist he had so much more spirit & resolution that he beat him till he declared he would fight no more, upon wch the schoolmaster's son bad him use him like a Coward, & rub his nose against the wall & accordingly Sʳ Isaac pulled him along by the ears & thrust his face against the side of the Church.

Not content with beating him physically, he insisted on worsting him academically as well; once on his way, he rose to be first in the school. As he rose, he left his trail behind him, his name carved on every bench he occupied. The benches do not survive, but a stone windowsill still bears one of his signatures.

By the time Stukeley was collecting anecdotes, Newton's genius was taken for granted. What everyone in Grantham remembered about him were "his strange inventions and extraordinary inclination for mechanical works." He filled his room in the garret of Clark's house with tools, spending all the money his mother gave him on them. While the other boys played their games, he made things from wood, not just doll furniture for the girls but also and especially models. A windmill was built north of Grantham while he was there. Although water wheels were common in the area, windmills were not, and the inhabitants of Grantham used to walk out to watch its construction for diversion. Only the schoolboy Newton inspected it so closely that he could build a model of it, as good a piece of workmanship as the original and one which worked when he set it on the roof. He went the original one better: He equipped his model with a treadmill run by a mouse which was urged on either by tugs on a string tied to its tail or by corn placed above it to the front. Newton called the mouse his miller. He made a little vehicle for himself. A four-wheeled cart run by a crank which he turned as he sat in it. He made a lantern of "crimpled paper" to light his way to school on dark winter mornings. The lantern, which he could simply fold up and put in his pocket for the day, had other possibilities; attached to the tail of a kite at night, it "wonderfully affrighted all the neighboring inhabitants for some time, and caus'd not a little discourse on market days, among the country people, when over their mugs of ale." By good fortune, Grantham was not burned to the ground.

Newton spent so much time at building that he frequently neglected his school work and fell behind, whereupon he turned to his books and quickly leaped ahead once more. Stokes remonstrated gently, but nothing could make him give up his mechanical contrivances. He could not leave them alone even on the Sabbath, although attending to them filled him with remorse. We know now that Newton found many of these contrivances in a book called *The Mysteries of Nature and Art* by John Bate. In another notebook from Grantham, with the information that he purchased it for 2½d in 1659, Newton took down extensive notes from Bate, on drawing, catching birds, making inks of various colors, and the like. Although they do not appear in his notes, most of his devices remembered in Grantham, including a windmill, were described in the book. Perhaps Newton's adolescent genius shrinks a little in the light of Bate's book. His genius is scarcely in doubt, however, and the fact is that he found a book which fed his natural interests. There is a touch of whimsy in some of these stories, wholly unexpected because wholly absent from the rest of his life. At this distance, there appears as well a pathetic attempt to ingratiate himself with his schoolfellows by such means. He made lanterns for them also, and who can doubt that they participated in the artificial meteor? When they flew kites, Newton investigated their properties to determine their ideal proportions and the best points to attach the strings. Apparently his efforts were in vain; he only convinced them of his greater ingenuity and thus completed their alienation. As Conduitt says, even when he played with the boys, he was always exercising his mind. Ordinary boys must have found him disconcerting. He told the Earl of Pembroke that the first experiment he ever made was on the day of Cromwell's death, when a great storm swept over England. By jumping first with the wind and then against it, and comparing his leaps with those of a calm day, he measured "the vis of the storm." When the boys were puzzled by his saying that the storm was a foot stronger than any he had known before, he showed them the marks that measured his leaps. According to one version of the story, he craftily used the wind to win a jumping contest – again the superior cunning which made him suspect.

There were also other recreations in Grantham. Among them were sundials. Apparently dials had attracted his attention even earlier; there is one mounted in the Colsterworth church supposedly cut by Newton when he was nine. Sundials involved much more than skill with tools; they presented an intellectual challenge. He filled poor Clark's house with dials – his own room, other rooms, the entry, wherever the sun came. He drove pegs into the walls to mark the hours, half-hours, and even quarterhours and tied strings with running balls to them to measure the shadows on successive days. By keeping a sort of almanac, he learned to distinguish the periods of the sun so that he could tell the equinoxes and solstices and even the days of the month. In the end the family and the neighbors came to consult "Isaac's dials." Thus did the majesty of the heavens and the uniformity of nature spread themselves unforgettably before him. According to Conduitt, he was still watching the sun at the end of his life. He observed the shadows in every room he frequented, and if asked, would look at the shadows instead of the clock to give the time.

He became proficient in drawing as well, and once more Clark's house bore the brunt of his enthusiasm. A later occupant of the garret room testified that the walls were covered with charcoal drawings of birds, beasts, men, ships, and plants. He also drew portraits of Charles I, John Donne, and the schoolmaster Stokes. A few circles and triangles also appeared on the walls – more of a forecast of the Newton we know than all of the portraits and birds and ships together. And on nearly every board, testifying to his identity like the desks in the school, stood the name "Isaac Newton," carved and therefore indelible.

What with carvings and drawings and sundials, pokings about in the shop, and peevish scenes over bread, the apothecary Clark may have looked forward to the departure of his precocious guest. That came late in 1659. Newton was turning seventeen. It was time that he face the realities of life and learn to manage his estate. With that end in view, his mother called him home to Woolsthorpe. From the beginning the attempt was a disaster. As the hero-worshipping Conduitt has it, his mind could not brook such "low employments." His mother appointed a trusty servant to teach him about the farm. Set to watch the sheep, he would build model water wheels in a brook, both overshot and undershot, with proper dams and sluices. The sheep meanwhile would stray into the neighbors' corn, and his mother would have to pay damages. The records on the manor court of Colsterworth show that on 28 October 1659 Newton was fined 3s 4d "for suffering his sheep to break ye stubbs on 23 ouf loes [loose? i.e., unenclosed] Furlongs," as well as 1s each on two other counts, "for suffering his swine to trespass in ye corn fields," and "for suffering his fence belonging to his yards to be out of repair." On market days, when he and the servant went to town to sell the produce of the farm and to purchase supplies, Newton would bribe the servant to drop him off beyond the first corner; he would spend the day building gadgets or with a book until the servant picked him up on the way home. If perhaps he went to town, he would run directly to his old room at Clark's where a stock of books awaited, and again the servant had to conduct the business. Going home to Woolsthorpe from Grantham, one had to mount Spittlegate hill immediately south of town. It was customary to dismount and lead one's horse up the steep hill. On one occasion, Newton became so lost in thought that he forgot to remount at the top and led the horse all the way home; on another occasion (or perhaps in another version of the same story) the horse slipped his bridle and went home while Isaac walked on, bridle in hand, unaware that the horse was gone. Apparently the servant stomached all of this. When Newton even forgot his meals, however, he despaired of ever teaching him.

Meanwhile, two other men were viewing Mrs. Smith's efforts from a different perspective. Her brother, the Reverened William Ayscough, had taken the young man's measure, and he urged his sister to send him back to school to prepare for the university. The schoolmaster, Mr. Stokes, was if anything more insistent. He remonstrated with Newton's mother on what a loss it was to bury such talent in rural pursuits, all the more so since the attempt was bound to fail. He even offered to remit the forty-shilling fee paid by boys not residents of Grantham, and he took Newton to board in his own home. Apparently Clark had had enough. In the autumn of 1660, as Charles II was learning to accustom himself to the perquisites of the throne, a more momentous event took place to the north. Isaac Newton returned to grammar school in Grantham, with the university in prospect beyond.

The evidence available indicates that the nine months at home were a nightmare. The list of sins of 1662 suggests constant tension: "Refusing to go to the close at my mothers command." "Striking many." "Peevishness with my mother." "With my sister." "Punching my sister." "Falling out with the servants." "Calling Derothy Rose a jade." He must have been insufferable. In Grantham, he had begun to sample how delicious learning could be. His inescapably intellectual nature had set him apart from the other boys, but he had no more been able to deny his nature to win their favor than a lion can give up his mane. Just as he had begun to commit himself to learning, however, he had been called back to the farm to spend his life herding sheep and shoveling dung. Everything within him rebelled against his fate, and fortune was on his side. By the intervention of Stokes and William Ayscough he was to feast on learning after all. His excitement still permeates Conduitt's account, blurred neither by sixty-five years nor by Conduitt's attempt at grandiloquence.

His genius now began to mount upwards apace & to shine out with more strength, & as he told me himself, he excelled particularly in making verses . . . . In everything he undertook he discovered an application equal to the pregnancy of his parts & exceeded the most sanguine expectations his master had conceived of him.

When Newton was ready finally to leave, Stokes set his favorite disciple before the school and with tears in his eyes made a speech in his praise, urging the others to follow his example. According to Stukeley, from whom Conduitt got the story, there were tears in the eyes of the other boys as well. We can imagine!

The schoolboys at Grantham were not the only group to whom Newton was a stranger and an enigma. To the servants at Woolsthorpe he was simply beyond comprehension. Surly on the one hand, inattentive on the other, not able even to remember his meals, he appeared both foolish and lazy in their eyes. They "rejoic'd at parting with him, declaring, he was fit for nothing but the 'Versity."

2

The Solitary Scholar

NEWTON SET OUT FOR CAMBRIDGE early in June 1661. There was no greater watershed in his life. Although he would return to Woolsthorpe infrequently during the next eighteen years, with two extended visits during the plague, spiritually he now left it, and what a later commentator has called the idiocy of rural life, once and for all. Three short years would put him beyond any possibility of return, though three more years, perhaps somewhat longer, had to pass before a permanent stay in Cambridge was assured. His accounts show that he stopped at Sewstern, presumably to check on his property there; and after spending a second night at Stilton as he skirted the Great Fens, he arrived at Cambridge on the fourth of June and presented himself at Trinity College the following day. If the procedures set forth in the statutes were followed, the senior dean and the head lecturer of the college examined him to determine if he was fit to hear lectures. He was admitted – although there is no record whatever of anything but the verdict, one feels constrained to add – "forthwith." He purchased a lock for his desk, a quart bottle and ink to fill it, a notebook, a pound of candles, and a chamber pot and was ready for whatever Cambridge might offer.

Admission to a college was not tantamount to admission to the university. Many delayed matriculation in the university; a considerable number who had no interest in a degree, to which alone matriculation was relevant, managed to avoid it altogether. Newton did intend to take a degree. On 8 July, together with a number of students recently admitted to Trinity and to other colleges, he duly swore that he would preserve the privileges of the university as much as in him lay, that he would save harmless its state, honor, and dignity as long as he lived, and that he would defend the same by his vote and counsel; and to testify to the same he paid his fee and saw his name entered in the university's matriculation book. He was now a full-fledged member of the university.

There is nothing surprising in the fact that Newton chose to enter Trinity, "the famousest College in the University" in the opinion of John Strype, the future ecclesiastical historian, who was an undergraduate in Jesus College at the time. As it happens, personal factors probably influenced Newton's choice more than the reputation of the college. The Reverend William Ayscough, his uncle, was a Trinity man, and according to the account that Conduitt later got from Mrs. Hatton, née Ayscough, the Reverend Mr. Ayscough persuaded Newton's mother to send him to Trinity. Stukeley heard in Grantham that Humphrey Babington, the brother of Mrs. Clark and fellow of Trinity, was responsible. The doctor, Stukeley wrote, "is said to have had a particular kindness for him, which probably was owing to his own ingenuity." There is some evidence to suggest a connection between Newton and Babington. "Mr Babingtons Woman," one of the bedmakers and chambermaids allowed to work in the college, appeared twice in the accounts Newton kept as a student, and he later indicated that he spent some of his time when he was home during the plague at neighboring Boothby Pagnell, where Babington held the rectory. As a fellow with considerable seniority such that in 1667 he became one of the eight senior fellows who, along with the master, controlled the college (and reaped its ripest rewards), and furthermore as a man who had demonstrated his access to royal favor with two letters mandate (i.e., orders from the king) immediately after the Restoration, Babington would be a powerful ally for a young man otherwise without connections. Both the nature of the college and the nature of Newton's studies made a powerful ally desirable at the least, and perhaps indispensable. For whatever reason, on 5 June 1661, the famousest college in the university, quite unaware, admitted its famousest student.

Newton entered Trinity as a subsizar, a poor student who earned his keep by performing menial tasks for the fellows, fellow commoners (very wealthy students who paid for privileges such as eating at high table – with the fellows of the college), and pensioners (the merely affluent). *Sizar* and *subsizar* were terms peculiar to Cambridge; the corresponding Oxonian word, *servitor*, expressed their position unambiguously. So did the statutes of Trinity College, which called them "scholares pauperes, qui nominentur Sizatores" and introduced the definition of their status by reference to the requirement laid on Christians to support paupers. The statutes allowed for thirteen sizars supported by the college, three to serve the master and ten for the ten fellows most senior; they also defined subsizars as students admitted in the same manner and subject to the same rules as sizars, but paying to hear lectures (at a rate lower than pensioners) and paying for their own food. That is, subsizars apparently were to be servants like sizars but not supported by the college – servants of fellows, of fellow commoners, and of pensioners, according to whatever arrangements they might make. Essentially identical in status, sizar and subsizar stood at the bottom of the Cambridge social structure, which repeated the distinctions of English society.

If all this was true, why was Newton a sizar? Only one possible answer presents itself. His mother, who had begrudged him further education in the first place, and (by one account) had sent him back to grammar school only when the forty-shilling fee was remitted, now begrudged him an allowance at the university that she could have afforded easily. Though her income probably exceeded £700 per annum, Newton's accounts seem to indicate that he received at most £10 per annum. There is a further possibility not inconsistent with the above. Newton may have gone to Trinity specifically as Humphrey Babington's sizar, perhaps to attend to the interests of Babington, who at that time was resident in Trinity only about four or five weeks a year. The payments mentioned above to "Mr Babingtons Woman" would fit into such a hypothesis. In the eighteenth century, the Ayscough family tradition recorded the story that "the pecuniary aid of some neighboring gentlemen" enabled Newton to study at Trinity. As the rector of Boothby Pagnell, Babington might fit that description. At a later time it does appear that Babington's support (that is, his influence, not his money) may have been crucial to Newton.

We cannot avoid a further question. What impact, if any, did his status as sizar have on Newton? He was, after all, heir to the lordship of a manor. If the manor itself was not grand, his family's economic status, thanks to the fortune of Barnabas Smith, ranked above that of most gentry. Newton was used to being served, not to serving. His own record, drawn up in 1662, indicates that he had used the servants at Woolsthorpe harshly, and they, for their part, had rejoiced to see him leave. It is hard to imagine that he did not find menial status galling. His status probably reinforced his natural propensity to isolation. Already in Grantham Newton had found it impossible to get along with his fellow students. If he thought he was escaping them to study with a superior breed in Cambridge, he was mistaken. The same boys were there; the names were all that differed. Only now he was their servant, carrying their bread and beer from the buttery and emptying their chamber pots.

The one surviving anecdote concerning his relations with other students suggests that the isolation and alienation of Grantham had traveled with Newton to Cambridge, intensified perhaps by his menial status. Well over half a century later, Nicholas Wickins, the son of Newton's chamber-fellow John Wickins, repeated what his father had told him about their meeting.

My Father's Intimacy with Him came by meer accident My Father's first Chamber-fellow being very disagreeable to him, he retired one day into ye Walks, where he found Mr Newton solitary and dejected; Upon entering into discourse they found their cause of Retiremt ye same, & thereupon agreed to shake off their present disorderly Companions & Chum together, wch they did as soon as conveniently they could, & so continued as long as my Father staid at College.

Because Wickins entered Trinity in January 1663, the encounter referred to occurred at least eighteen months after Newton's admission. I am inclined to think that the walks of Trinity had frequently known a solitary figure during those eighteen months, as they would for thirty-five years more. With the exception of Wickins, Newton formed no single friendship from among his fellow students that played a perceptible role in his life, though he would live on in Trinity with some of them until 1696, and even his relation with Wickins was ambiguous. Correspondingly, when Newton became England's most famous philosopher, none of his fellow students left any recorded mention that they had once known him. The sober, silent, thinking lad of Grantham had become the solitary and dejected scholar of Cambridge.

Significantly, I think, Wickins was a pensioner. Trinity was less rigid in segregating sizars than some of the colleges. It did not prescribe separate academic gowns for them, and the possibility of a sizar's "chumming" (that is, sharing a chamber) with a pensioner existed. At first blush, it might appear that Newton was more apt to find congenial companions among the other sizars. By and large, they were the serious students. Whereas only 30 percent of the gentlemen who entered Cambridge continued to the degree, roughly four out of five sizars commenced B.A. On the whole, however, they were a plodding group, narrowly vocational in outlook, lower-class youths grimly intent on ecclesiastical preferment as the means to advancement. Because he had entered Trinity at eighteen, Newton was at least one year older than the average and perhaps two, another factor that separated him from them. Genius of Newton's order does not readily find companionship in any society in any age. He was perhaps even less apt to find it among the sizars of Restoration Cambridge. As in Grantham, he was unable to conceal his brilliance. "When he was young & first at university," his niece Catherine Conduitt told her husband, "he played at drafts & if any gave him first move sure to beat them."

In the summer of 1662, Newton underwent some sort of religious crisis. At least he felt impelled to examine the state of his conscience at Whitsunday, to draw up a list of his sins before that date, and to start a list of those committed thereafter. His earnestness did not survive long enough to extend the second list very far. Lest it fall under the wrong eyes, he recorded his sins in cipher, using Shelton's system of shortwriting just as Samuel Pepys was using it at the same time for a livelier and more revealing record. Many of the incidents that Newton remembered with shame belonged to Grantham and to Woolsthorpe, but some of them belonged to Cambridge: "Having uncleane thoughts words and actions and dreamese." He had not kept the Lord's day as he ought: "Making pies on Sunday night"; "Squirting water on Thy day"; "Swimming in a kimnel [a tub] on Thy day"; "Idle discourse on Thy day and at othertimes"; "Carlessly hearing and committing many sermons." He had not loved the Lord his God with all his heart and with all his soul and with all his mind: "Setting my heart on money learning pleasure more than Thee"; "Not turning nearer to Thee for my affections"; "Not living according to my belief"; "Not loving Thee for Thy self"; "Not desiring Thy ordinances"; "Not fearing Thee so as not to offend Thee"; "Fearing man above Thee"; "Neglecting to pray." Relying upon this confession and upon his interpretation of the lists of words in the Morgan notebook, Professor Frank Manuel concludes that Newton was borne down "by a sense of guilt and by doubt and self-denigration. The scrupulosity, punitiveness, austerity, discipline, and industriousness of a morality that may be called puritanical for lack of a better word were early stamped upon his character. He had a built-in censor and lived ever under the Taskmaster's eye." Newton's undergraduate expenditures appear to bear out Manuel's judgment. If he treated himself now and then to cherries, "marmolet," custards, and even a little wine on occasion, he felt obliged to enter them under *Otiosi et frustra expensa* as opposed to *Impensa propria*, which included clothes, books, and academic supplies. He even considered beer and ale as *otiosi*, though we might judge them *propria* as we reflect on the water available.

୨♠

Meanwhile, along with the problems of daily life there were also studies. By 1661 the official curriculum of Cambridge, prescribed by statute nearly a century before, was in an advanced state of decomposition. Study at Cambridge had not seriously broken the mold in which it had been cast four centuries earlier, with its focus centered on Aristotle. When it had been formulated initially, it had embodied the most advanced position of European philosophy. By 1661, European philosophy had moved on, and academic Aristotelianism represented an intellectual backwater maintained in part by the legal mandate of a curriculum enacted as law and in part by men who had a vested interest in continuing a system to which they had bound their lives. Intellectual vigor had departed long since. It was becoming an exercise performed by rote, without enthusiasm.

One of Newton's first purchases in Cambridge was a notebook, and probably it was in this one that he entered the fruits of his reading in the established curriculum. In fact, he did not finish any of the books that he started. He had found other reading. Perhaps one should not regard history as alternative reading; it figured strongly in some of the programs of study that tutors prescribed. At any rate, two history books, Hall's *Chronicles* and Sleidan's *Four Monarchies*, were among his early purchases in Cambridge. Although he left nothing among his undergraduate notes from his reading in these works, chronology remained, in close association with his study of the prophecies, one of his abiding interests. For a brief time, about 1663, he examined judicial astrology, according to a conversation he had with Conduitt near the end of his life. Astrology was never part of the curriculum. Phonetics and a universal philosophical language also had nothing to do with established studies, although they, or at least the idea of a universal language, were live centers of intellectual interest at the time. There had been a number of schemes for a universal language based, as Newton expressed it, on "ye natures of things themselves wch is ye same to all Nations." Sometime during his undergraduate career, Newton came across this literature; he drew especially upon George Dalgarno's *Ars signorum* (1661). To it he attached an interest in phonetics which may have derived from his study of Shelton's system of shortwriting. Other interests soon pushed the universal language aside, and he never returned to it.

Frequently, as in the case of John Wilkins's *Essay Toward a Real Character and a Philosophic Language* (published in 1668, after Newton's venture in this field), the concept of a universal language was coupled with criticism of Aristotelian philosophy, which was held not to express the "real" nature of things. Newton's youthful exercise was not so. Couched in Aristotelian terms, it reflected the sole philosophy to which he had been introduced. Such was not long the case, however. In the notebook in which he had entered the fruits of his study advancing from each end, about a hundred pages remained empty in the center. Two pages devoted to Descartes's metaphysics bluntly interrupted the Aristotelianism of the texts he had been reading. A few pages further on he entered the title "Questiones quaedam Philosophcae" and laid out a set of headings under which to collect the notes from a new course of readings. Somewhat later, he wrote a slogan over the title "Amicus Plato amicus Aristoteles magis amica veritas." Whatever there may be of truth in the pages that follow, certainly there is nothing from Plato or Aristotle. Notes from Descartes, whose works Newton thoroughly digested in a way that he never had those of Aristotle, appear throughout the "Quaestiones." Nor had he confined himself to Descartes. He had also read Walter Charleton's English epitome and translation of Pierre Gassendi, and perhaps some of Gassendi as well. He had read Galileo's *Dialogue,* though apparently not his *Discourses.* He had read Robert Boyle, Thomas Hobbes, Kenelm Digby, Joseph Glanville, Henry More, and no doubt others as well. Veritas, Newton's new friend, was none other than *philosophia mechanica.*

There is no way conclusively to date the beginning of the "Quaestiones" although various considerations suggest some time not too late in 1664. Equally there is no way to state with assurance the agency involved, but everything we know about Cambridge suggests it had little to do as an institution with leading Newton to the new philosophy. One piece of testimony indicates that Descartes was very much in the air at the time, so the advice of a tutor would scarcely have been required. Roger North, an undergraduate in Cambridge in 1667–8, whose tutor, his brother, did not wish to be bothered and left him to follow his own inclinations, "found such a stir about Descartes, some railing at him and forbidding the reading him as if he had impugned the very Gospel. And yet there was a general inclination, especially of the brisk part of the University, to use him . . . ." Newton's notes imply that he also found a stir about Descartes and decided to investigate him. Beyond Descartes, we are left wholly to speculation, but it is not hard to imagine the process whereby Newton was led on from one author to another into a totally new world of thought. At last he had found what he had come seeking in Cambridge. Without hesitation, he embraced it as his own. The very laxity of the university now worked to his advantage. His tutor, Benjamin Pulleyn, was probably happy enough not to be bothered, and Newton could pursue his interest unhindered.

He set down forty-five headings under which to organize the fruits of his reading, beginning with general topics on the nature of matter, place, time, and motion, proceeding to the cosmic order, then to a large number of tactile qualities (such as rarity, fluidity, softness), followed by questions on violent motion, occult qualities, light, colors, vision, sensation in general, and finally concluding with a set of miscellaneous topics not all of which appear to have been in the initial list. Under some of the headings he never entered anything; under others he found so much that he had to continue the entries elsewhere. The title "Quaestiones" adequately describes the whole in that the tone was one of constant questioning. The questions were posed within certain limits, however. They probed details of the mechanical philosophy; they did not question the philosophy as a whole. Newton had left the world of Aristotle forever.

One product of his new world view was a temporary interest in perpetual motion. The mechanical philosophy pictured a world in constant flux. Newton, the tinkerer from Grantham, thought of various devices, in effect windmills and water wheels, to tap the currents of invisible matter. For example, he adopted the view that gravity (heaviness) is caused by the descent of a subtle invisible matter which strikes all bodies and carries them down. "Whither ye rays of gravity may bee stopped by reflecting or refracting y^m, if so a perpetuall motion may bee made one of these two ways." He drew sketches of mill-like devices that the stream of invisible matter would turn. Under the heading of magnetism, he proposed analogous devices.

Most of the entries in the "Quaestiones" were derivative, notes from Newton's reading. Nevertheless, the whole carried the unmistakable imprint of its author. To a remarkable degree the "Quaestiones" foreshadowed the problems on which his career in science would focus and the method by which he would attack them. As to the latter, the title "Quaestiones," which describes not only the set of headings but also their content, suggests the active questioning that lay behind Newton's procedure of experimental enquiry. Many of the questions were directed to the authors he was reading, whose opinions he did not merely register passively. Descartes's theory of light raised a number of objections.

Light cannot be by pression &[c] for y^n wee should see in the night a wel or better y^n in ye day we should se a bright light above us becaus we are pressed downewards . . . there could be no refraction since ye same matter cannot presse 2 ways. a little body interposed could not hinder us from seing pression could not render shapes so distinct. ye sun could not be quite eclipsed ye Moone & planetts would shine like sunns. A man goeing or running would see in ye night. When a fire or candle is extinguish we lookeing another way should see a light. The whole East would shine in ye day time & ye west in ye night by reason of ye flood wch carrys o^r Vortex a light would shine from ye Earth since ye subtill matter tends from ye center. There is ye greatest pression on y^t side of ye earth from y^e ⊙ [sun] or else it would not move about in equilibrio but from y^e ⊙, therefore ye nights should be lightest.

These were very searching questions indeed directed to the Cartesian explanation of light. Under the heading "Of ye Celestiall matter & orbes" he added a few more, pointing out that eclipses would be impossible according to the Cartesian theory because solid bodies could transmit the pressure in the vortex as well as the fluid matter of the heavens. Every statement in these passages was an implicit experiment, an observation of a critical phenomenon that ought to appear if the theory were true. When he considered theories of colors, he proceeded in the same way. Do colors arise from mixtures of darkness and light? If they do, a printed page, black letters on a white sheet, ought to appear colored at a distance – another implicit experiment. Some of the experiments were posed explicitly. Descartes had referred the tides to the pressure of the moon on the fluid matter of the tiny vortex surrounding the earth. In a work by Boyle, Newton found a proposal to test the theory by correlating tides with the readings of barometers, which ought to register the same pressure. Immediately he began to think of other consequences the theory should entail.

Observe if yᵉ sea water rise not in days & fall at nights by reason of yᵉ earth pressing from ☉ uppon yᵉ night water &c. Try also whither yᵉ water is higher in mornings or evenings to know whither ⊖ [earth] or its vortex press forward most in its annuall motion . . . . Try whither yᵉ seas flux & reflux bee greater in Spring or Autume in winter or Sommer by reason of yᵉ ⊖s Aphelion & perihelion. Whither yᵉ Earth moved out of its Vortexes center bye Moones pression cause not a monethly Parallax in Mars. &c.

There was no suggestion that Newton had made any of these observations. Nevertheless, if the essence of experimental procedure is active questioning whereby consequences that ought to follow from a theory are put to the test, Newton the experimental scientist was born with the "Quaestiones." In 1664, such a method of inquiry had been little used. Newton's example was to be a powerful factor in helping experimental procedure convert natural philosophy into natural science.

As he became interested in light and vision, for which some forms of experimentation required no equipment beyond his own eyes, Newton plunged forward with little thought of the consequences. To test the power of fantasy, he looked at the sun with one eye until all pale bodies seen with that eye appeared red and dark ones blue. After "yᵉ motion of yᵉ spirits in my eye were almost decayed" so that things were beginning to appear normal, he closed his eye and "heightned [his] fantasie" of seeing the sun. Spots of various hues appeared to his eye, and when he opened it again pale bodies appeared red and dark ones blue as though he had been looking at the sun. He concluded that his fantasy was able to excite the spirits in his optic nerve quite as well as the sun. He also came close to ruining his eyes, and he had to shut himself up in the dark for several days before he could rid himself of the fantasies of color. Newton left the sun alone after that, but not his eyes. A year or so later when he was developing his theory of colors he slipped a bodkin "betwixt my eye & yᵉ bone as neare to yᵉ backside of my eye as I could" in order to alter the curvature of the retina and to observe the colored circles that appeared as he pressed.

How did he fail to blind himself? In the grip of discovery, Newton did not pause to reckon the cost.

The content of the "Quaestiones" is equally redolent of the future Newton. The passages "Of Motion" and especially "Of violent Motion" mark his introduction to the science of mechanics. The latter passage, really an essay, attacked the Aristotelian explanation of projectile motion and concluded that the continued motion of a projectile after it separates from the projector is due to its "naturall gravity." By "gravity" he referred in this instance to an atomistic doctrine that every atom has an inherent motility, called gravity, by which it moves. The doctrine was similar, though by no means identical, to the medieval theory of impetus, which struggled with the principle of inertia for Newton's allegiance for twenty years. He considered the cosmic order and Descartes's system of vortices. Elsewhere in the notebook, in a hand that corresponds with the later entries in the "Quaestiones," Newton also took notes from Thomas Streete's *Astronomia carolina*, which effectively introduced him to Keplerian astronomy. He pondered the cause of gravity (that is, heaviness) and pointed out that the "matter" which causes bodies to fall must act on their innermost particles and not merely on their surfaces. As I have already stated, light and colors occupied a considerable portion of the "Quaestiones"; in their pages Newton recorded the central insight to the demonstration of which his entire work in optics was directed, that ordinary light from the sun is heterogeneous, and that phenomena of colors arise, not from the modification of homogeneous light, as prevailing theory had it, but from the separation or analysis of the heterogeneous mixture into its components.

Although the stance of questioning remained predominant in the "Quaestiones," an inchoate natural philosophy beginning to take shape can be dimly perceived. If Descartes was cited most frequently, his influence did not in the end dominate the "Quaestiones." Two other systems challenged his authority. On the one hand, Gassendi's atomistic philosophy, known to Newton at this time primarily through Charleton's *Physiologia*, offered a rival mechanical system. More than anything else, the "Quaestiones" were a dialogue in which Newton weighed the virtues of the two systems. Although he appeared to reach no final verdict, it is clear that he inclined already toward atomism. After deploying the standard arguments against a plenum, Newton opted for atoms, though not, or at least not initially, Gassendi's atoms. I have already quoted Newton's objections to Descartes's conception of light and his explanation of the tides, and I have indicated that he accepted a different view of the cause of gravity (heaviness). Matter and light were the most important; to reject Descartes's opinions on these two issues was to shatter the cohesion of his natural philosophy beyond hope of repair. In his discussions of light and color, Newton left no doubt that he held the corpuscular conception. Descartes may have introduced him to the mechanical philosophy, but Newton quickly transferred his allegiance to atomism.

There is also the possibility that the writings of Henry More guided Newton into the mechanical philosophy. Descartes's name appeared so frequently in them that he could not have failed to notice it. Whichever he came to first, More represented the other current of thought that tempered Newton's enthusiasm for Descartes. More's views exerted a strong influence on the original essay on atoms that Newton wrote in the "Quaestiones." Newton later crossed the essay out, however, and it was not here that More's position was vital. Like the other Cambridge Platonists, Henry More was concerned by the mechanical philosophy's possible exclusion of God and spirit from the operation of physical nature. Whereas initially he welcomed Descartes as an ally of religion, the more he contemplated his sytem of nature, the more its implications alarmed him. In Hobbes he saw the dangers spelled out explicitly. More was concerned to reinstall spirit in the continuing operation of nature, all of nature. Especially in the last four entries of the "Quaestiones" – "Of God," "Of ye Creation," "Of ye soule," and "Of Sleepe and Dreams &c," which appear by their position to be later additions to the original set of headings – similar concerns made a tentative appearance in the "Quaestiones." Their role in Newton's thought was destined to grow, diluting and modifying his initial mechanistic views.

Meanwhile, natural philosophy was not the only new study Newton discovered. He found mathematics as well. As with natural philosophy, we have Newton's original notes that chart his course. We also have a number of accounts, several in Newton's own words of which one from 1699 is the most important, one in Conduitt's memorandum of a conversation with Newton on 31 August 1726, and another in a memorandum of November 1727, soon after Newton's death, by Abraham DeMoivre. The earliest of them dated from thirty-five years after the events it described. Nevertheless, a reasonably consistent account, which is also reasonably consistent with Newton's reading notes, emerges from them.

July 4th 1699. By consulting an accompt of my expenses at Cambridge in the years 1663 and 1664 [Newton wrote as he looked over some early notes] I find that in ye year 1664 a little before Christmas I being then senior Sophister, I bought Schooten's Miscellanies & Cartes's Geometry (having read this Geometry & Oughtred's Clavis above half a year before) & borrowed Wallis's works & by consequence made these Annotations out of Schooten & Wallis in winter between the years 1664 & 1665. At wch time I found the method of Infinite series. And in summer 1665 being forced from Cambridge by the Plague I computed ye area of ye Hyperbola at Boothby in Lincolnshire to two & fifty figures by the same method.

In Conduitt's memorandum, it all began when Newton lit upon some books on judicial astrology (an event which DeMoivre placed at Sturbridge Fair in 1663). Being unable to cast a figure, he bought a copy of Euclid and used the index to locate the two or three theorems he needed; when he found them obvious, "he despised that as a trifling book . . . ." DeMoivre's account agreed with Conduitt's except that he had Newton go on in Euclid to more difficult propositions, such as the Pythagorean theorem, whereupon he changed his opinion, and read all of Euclid through twice. Such early study of Euclid does not agree either with Newton's notes or with other parts of what he told Conduitt. Pemberton also recorded Newton's regret that he had not given more attention to Euclid before he applied himself to Descartes.

He bought Descartes's Geometry & read it by himself [Conduitt continued in language very similar to that in the DeMoivre account] when he was got over 2 or 3 pages he could understand no farther than he began again & got 3 or 4 pages farther till he came to another difficult place, than he began again & advanced farther & continued so doing till he made himself Master of the whole without having the least light or instruction from any body.

Both accounts agree in making Newton an autodidact in mathematics, as he was in natural philosophy. Nearly twenty years later, when he was recommending Edward Paget for the position of mathematical master at Christ's Hospital, Newton probably had his own experience in mind as he specified Paget's qualifications. Paget understood the several branches of mathematics, he said, "& wch is ye surest character of a true Mathematicall Genius, learned these of his owne inclination, & by his owne industry without a Teacher."

There had been even less mathematics in the university than natural philosophy; not surprisingly, no stories survive of undergraduates being stirred by Descartes's *Geometry*. Nevertheless, there is a curious coincidence of time, which has generally been ignored. The Lucasian chair of mathematics, which Newton himself would soon occupy, was established in 1663, and the first professor, Isaac Barrow, delivered his inaugural series of lectures in 1664, beginning on 14 March. Contrary to frequent assertions, Barrow was not Newton's tutor, and there is no evidence of any familiarity between them at this time. At least twice, Newton implied that he had attended the lectures, however; and though they would probably not have directed him to Descartes, given Barrow's mathematical predilections, and though Barrow was not a major influence on him, they could have stimulated his interest in mathematics. One wonders as well who in Cambridge could have loaned him a copy of Wallis if it was not Barrow. In any event, the coincidence in time is so close that it strains credulity to deny any connection between the lectures and Newton's sudden interest.

Newton's own notes agree with the accounts of Conduitt and DeMoivre that he plunged straight into modern analysis with no appreciable background in classical geometry. They agree as well with the centrality given Descartes. Franz van Schooten's second Latin edition of the *Geometry*, with its wealth of additional commentaries, was his basic text, supplemented by Schooten's *Miscellanies*, Viète's works, Oughtred's algebra (the *Clavis* Newton mentioned), and Wallis's *Arithmetica infinitorum*. In roughly a year, without benefit of instruction, he mastered the entire achievement of seventeenth-century analysis and began to break new ground.

&.

Newton's surrender to his new studies was not without danger. In order to pursue them to a fruitful conclusion, he had to win a permanent position in Cambridge, but rewards in Cambridge were not being passed out for excellence in mathematics and mechanical philosophy. Fellowships in Trinity went only to those who had first been elected as an undergraduate to one of the sixty-two scholarships supported by the college. In his first three years, Newton had not distinguished himself in any way. Trinity had twenty-one exhibitions, which carried annual stipends of about four pounds each. The college records give no indication of the criteria of selection. It is difficult to imagine that academic promise did not figure, though need may have been the decisive factor. Suffice it to say that Newton did not appear among the ten, nearly all Pulleyn's pupils, who received exhibitions in 1662 and 1663.

Many features of the college worked to lessen his chances of a scholarship. Statistics indicate that sizars had less chance than pensioners, especially when enrollment was up and demand high, as they were in the 1660s. Influence and connections were essential characteristics of the system of patronage that impinged on the whole university to the injury of those, sizars above all, who lacked sponsors in high places. Newton's chances were further diminished by the privileged group of Westminster scholars who automatically received at least a third of the scholarships year after year, and with the scholarships the top rungs on the ladder of seniority for their year. During the entire century, a good half of Trinity's fellows came from Westminster School, and roughly that proportion held for the large group elected scholars in 1664. Indeed, with 1664 Newton faced a crisis. Trinity held elections to scholarships only every three or four years. The election in 1664 was the only one during his career as a student. If Newton were not elected then, all hope of permanent residence in Cambridge would vanish forever. He chose exactly that time to throw over the recognized studies and pursue a course that had no standing whatever in the college's scheme of values.

Perhaps the approaching elections, to be held in April, with their attendant examinations explain an otherwise anomalous feature of Newton's notes on the established curriculum. Having dropped Magirus's peripatetic *Physics*, he took it up again and plowed his way through two more chapters. Likewise, he started Vossius's *Rhetoric* and the *Ethics* of Eustacius of St. Paul about this time – and likewise failed to finish both works. In the three cases, the notes suggest last-minute boning for an examination. Newton's own account, as related to Conduitt, implies that his tutor Pulleyn may have recognized his pupil's brilliance and tried to help him by enlisting Isaac Barrow, the one man in Trinity fit to judge his competence in the unorthodox studies he had undertaken. The gesture nearly capped the debacle, since Newton had been unorthodox even in his unorthodoxy.

When he stood to be scholar of the house his tutour sent him to Dr Barrow then Mathematical professor to be examined, the D^r examined him in Euclid wch Sr I. had neglected & knew little or nothing of, & never asked him about Descartes's Geometry wch he was master of Sr I. was too modest to mention it himself & Dr Barrow could not imagine that any one could have read that book without being first master of Euclid, so that Dr Barrow conceived then but an indifferent opinion of him but however he was made scholar of the house.

The final clause is true; on 28 April 1664 Newton was elected to a scholarship. It also poses a quandary: What can explain the decision? Perhaps the explanation is the obvious one that springs immediately to mind. Newton's genius readily outshone the mediocrity around him even in studies he had abandoned. Such an explanation seems to conflict, however, with Newton's account of the impression he made on Barrow, the leading intellect of the college. Moreover, the realities of Cambridge in 1664 suggest another explanation, that Newton had a powerful advocate within the college. There is good reason to think he had such an advocate. In 1669, as a new fellow, he was appointed tutor of a fellow commoner. Tutoring fellow commoners was lucrative business usually reserved for important fellows. Two candidates for Newton's patron present themselves. One is Barrow himself, despite the story. It is not impossible that Newton was misled as to the impression he made. That is speculation, however. It is not speculation that in 1668–9, Barrow was familiar enough with Newton's work to send him Mercator's *Logarithmotechnia* when he saw that it seemed to forestall some of his work. In 1669, he obtained the Lucasian chair for Newton when he himself resigned, and in 1675, Barrow appears to have been decisive in obtaining a royal dispensation for Newton. The other and more likely candidate is Humphrey Babington. Recall Newton's accounts, which show that he employed "Mr Babingtons Woman." Recall his statement that during the plague he was, at least part of the time, at Boothby, not far from Woolsthorpe, where the same Mr. Babington was rector. Mr. Babington was also the brother of Mrs. Clark, with whom Newton had lodged in Grantham. Most important of all, he was approaching the status of a senior fellow, one of the eight fellows at the top of the ladder of seniority, who ran the college in conjunction with the master. Moreover, the college would not have forgotten that he stood well with the king; twice he had obtained letters mandate in his favor in recent years. When Babington was later bursar of the college, Newton drew up tables to aid him in renewing college leases, and the two continued to be associated in various academic affairs until Babington's death. Because Babington was resident only four or five weeks of the year at this time, however, his opportunity to influence the election may have been small. Four years earlier, the Reverend William Ayscough and Mr. Stokes had rescued Newton from rural oblivion. Someone performed that service again in April 1664, and on the whole Humphrey Babington appears most likely to have been the one.

With his election, Newton ceased to be a sizar. He now received commons from the college, a livery allowance of 13s 4d per year, and a stipend of the same amount. Far more important, he received the assurance of at least four more years of unconstrained study, until 1668 when he would incept M.A., with the possibility of indefinite extension should he obtain a fellowship. The threat had lifted. He could abandon himself completely to the studies he had found. The capacity Newton had shown as a schoolboy for ecstasy, total surrender to a commanding interest, now found in his early manhood its mature intellectual manifestation. The tentativeness suggested by the earlier unfinished notes vanished, to be replaced by the passionate study of a man possessed. Such was the characteristic that his chamber-fellow Wickins remembered, having observed it no doubt at the time with the total incomprehension of the Woolsthorpe servants. Once at work on a problem, he would forget his meals. His cat grew very fat on the food he left standing on his tray. (No peculiarity of Newton's amazed his contemporaries more consistently; clearly food was not something they trifled with.) He would forget to sleep, and Wickins would find him the next morning, satisfied with having discovered some proposition and wholly unconcerned with the night's sleep he had lost. "He sate up so often long in the year 1664 to observe a comet that appeared then," Newton told Conduitt, "that he found himself much disordered and learned from thence to go to bed betimes." Part of the story is true; he entered his observations of the comet into the "Quaestiones." The rest of it is patently false, as Conduitt knew from personal experience. Newton never learned to go to bed betimes once a problem seized him. Even when he was an old man the servants had to call him to dinner half an hour before it was ready, and when he came down, if he chanced to see a book or a paper, he would let his dinner stand for hours. He ate the gruel or milk with eggs prepared for his supper cold for breakfast. Conduitt observed Newton long after his years of creativity. The tension of the quest that consumed him in 1664 and the years that followed stretched whatever neuroses he had brought from Woolsthorpe to their utmost limits. He was "much disordered" more than once, and not only from observing comets.

His discovery of the new analysis and the new natural philosophy in 1664 marked the beginning of Newton's scientific career. He considered the "Quaestiones" important enough that he later composed an index to them to supplement their initial organization under topics. Sailing away from the old world of academic Aristotelianism, Newton launched his voyage toward the new. The passage was swift.

3

Anni Mirabiles

MORE THAN ANYTHING ELSE mathematics dominated Newton's attention during the months that followed his discovery of the new world of science, although it did not completely obliterate other interests. Sometime during this period he also found time to compose the "Quaestiones," in which he digested current natural philosophy as efficiently as he did mathematics. The other mathematicians and natural philosophers of Europe were unaware that a young man named Isaac Newton even existed. To those who knew of him, his fellow students in Trinity, he was an enigma. The first blossoms of his genius flowered in private, observed silently by his own eyes alone in the years 1664 to 1666, his *anni mirabiles*.

In addition to mathematics and natural philosophy, the university also made certain demands on his time and attention. He was scheduled to commence Bachelor of Arts in 1665, and regulations demanded that he devote the Lent term to the practice of standing *in quadragesima*. Pictured in our imagination, the scene has a surrealistic quality, medieval disputations juxtaposed with the birth pains of the calculus. An investigation of curvature was dated 20 February 1665, in the middle of the quadragesimal exercises, and in his various accounts of his mathematical development he assigned the binomial expansion to the winter between 1664 and 1665. While Stukeley was a student at Cambridge more than thirty years later, he heard that when Newton stood for his B.A. degree "he was put to second posing, or lost his groats as they term it, which is look'd upon as disgraceful." The story raises several problems. The senate had already passed the grace granting his degree before the exercises were held, and Newton signed for this degree with the other candidates. If the story has any substance, it has to apply to prior examinations in the college. Nevertheless, as Stukeley remarked, it does not seem strange because Newton was not much concerned with the standard curriculum. Once more the laxity of the university worked to his advantage. Newton commenced B.A. largely because the university no longer believed in its own curriculum with enough conviction to enforce it.

In the summer of 1665, a disaster descended on many parts of England, including Cambridge. It had "pleased Almighty God in his just severity," as Emmanuel College put it, "to visit this towne of Cambridge with the plague of pestilence." Although Cambridge could not know it and did little in the following years to appease divine severity, the two-year visitation was the last time God would choose to chastise it in this manner. On 1 September, the city government canceled Sturbridge Fair and prohibited all public meetings. On 10 October, the senate of the university discontinued sermons at Great St. Mary's and exercises in the public schools. In fact, the colleges had packed up and dispersed long before. Trinity recorded a conclusion on 7 August that "all Fellows & Scholars which now go into the Country upon occasion of the Pestilence shall be allowed ye usuall Rates for their Commons for ye space of ye month following." The records of the steward make it clear that the college, though ahead of the university, was behind many of its residents, who had fled already and therefore collected no allowance for the last month of the summer quarter. For eight months the university was nearly deserted. In the middle of March when no deaths had been reported for six weeks, the university invited its fellows and students to return. By June it was evident that the visitation was not concluded. A second exodus occurred, and the university was able to resume in earnest only in the spring of 1667.

Many of the students attempted to continue organized study by moving with their tutors to some neighboring village. Because Newton was entirely independent in his studies and had had his independence confirmed with a recent B.A., he found no occasion to follow Benjamin Pulleyn. He returned instead to Woolsthorpe. He must have left before 7 August 1665 because he did not receive the extra allowance granted on that date. His accounts show that he returned on 20 March 1666. He received the standard extra commons in 1666 and hence probably left for home in June. His accounts show again that he returned in 1667, late in April.

Much has been made of the plague years in Newton's life. He mentioned them in his account of his mathematics. The story of the apple, set in the country, implies the stay in Woolsthorpe. In another much-quoted statement written in connection with the calculus controversy about fifty years later, Newton mentioned the plague years again.

In the beginning of the year 1665 I found the Method of approximating series & the Rule for reducing any dignity of any Binomial into such a series. The same year in May I found the method of Tangents of Gregory & Slusius, & in November had the direct method of fluxions & the next year in January had the Theory of Colours & in May following I had entrance into ye inverse method of fluxions. And the same year I began to think of gravity extending to ye orb of the Moon & (having found out how to estimate the force with w^{ch} [a] globe revolving within a sphere presses the surface of the sphere) from Keplers rule of the periodical times of the Planets being in sesquialterate proportion of their distances from the center of their Orbs, I deduced that the forces wch keep the Planets in their Orbs must [be] reciprocally as the squares of their distances from the centers about wch they revolve: & thereby compared the force requisite to keep the Moon in her Orb with the force of gravity at the surface of the earth, & found them answer pretty nearly. All this was in the two plague years of 1665-1666. For in those days I was in the prime of my age for invention & minded Mathematicks & Philosophy more then at any time since.

From this statement, combined with the other statements about his mathematics and the story of the apple, has come the myth of an *annus mirabilis* associated with Woolsthorpe. From one point of view, the leisure of his forced vacation from academic requirements gave him time to reflect. From another point of view, his return to the maternal bosom provided a crucial psychological stimulus. Either theory is impossible to prove or to disprove. We may be moderately skeptical of the second as we recall the less than total bliss of his year at home in 1660. It may be relevant as well that his last act before he returned to Cambridge was to pry an extra £10 from the tight fist of his mother. In any event, exclusive attention to the plague years and Woolsthorpe disregards the continuity of his development. Intellectually, Newton departed from Cambridge more than a year before the plague drove him away physically. He took important steps toward the calculus in the spring of 1665, before the plague struck, and he wrote two important papers during May in 1666 while he was back. Similarly, his development as a physicist flowed without break from the "Quaestiones quaedam Philosophicae." If we focus our attention on the record of his studies, the plague and Woolsthorpe fade in importance in comparison to the continuity of his growth. The year 1666 was no more *mirabilis* than 1665 and 1664. The miracle lay in the incredible program of study undertaken in private and prosecuted alone by a young man who thereby assimilated the achievement of a century and placed himself at the forefront of European mathematics and science.

ра

Looking back from the beginning of 1666, one finds it difficult to believe that Newton touched anything but mathematics during the preceding eighteen months. In his age of celebrity, Newton was asked how he had discovered the law of universal gravitation. "By thinking on it continually" was the reply. No better characterization of the man can be given, not only in its delineation of a life whose central adventure lay in the world of thought rather than action but also in its description of his mode of work. Seen from afar, Newton's intellectual life appears unimaginably rich. He embraced nothing less than the whole of natural philosophy, which he explored from several vantage points, ranging all the way from mathematical physics to alchemy. Within natural philosophy, he gave new direction to optics, mechanics, and celestial dynamics, and he invented the mathematical tool that has enabled modern science further to explore the paths he first blazed. He sought as well to plumb the mind of God and His eternal plan for the world and humankind as it was presented in the biblical prophecies. When we examine Newton's grandiose adventure minutely, it turns out to be a mixture of discrete pieces rather than a homogeneous mélange. His career was episodic. What he thought on, he thought on continually, which is to say exclusively, or nearly exclusively. What seized his attention in 1664, to the virtual exclusion of everything else, was mathematics.

John Conduitt, the husband of Newton's niece and his intended biographer, almost invariably smothered whatever insight he had in a froth of grandiloquence. One of his figures, applied to Newton's early career, bears repetition, however: "he began with the most crabbed studies (like a high spirited horse who must be first broke in plowed grounds & the roughest & steepest ways or could otherwise be kept within no bounds)." Newton was to voyage over many strange seas of thought, speculative adventures from which more than one explorer of the seventeenth century never returned. The discipline that mathematics imposed on his fertile imagination marked the difference between wild tacks of fancy and fruitful discovery. It was supremely important that, almost first, mathematics commanded his attention.

The surviving notes of his initial studies in mathematics bear out the various anecdotes that he plunged straightforward into Descartes's *Geometry* and modern analysis. The time was almost certainly 1664, probably in the spring or summer. His primary vehicle was Schooten's pivotal second Latin edition of Descartes's *Geometry* with its wealth of additional commentaries, supported by reading in algebra, especially from the works of Viète. He also made early contact with the mathematics of infinitesimals as represented by John Wallis. It is quite impossible to determine from the notes which came first. It is equally impossible to see that anything important hinges on their chronological order. What matters is the voracity with which he devoured whatever mathematics he found. William Whiston later remarked that in mathematics Newton "could sometimes see almost by Intuition, even without Demonstration . . . ." Whiston had a proposition in the *Principia* in mind, but an examination of Newton's self-education in mathematics compels one to a similar judgment. Within six months of his initiation into mathematics, some of his reading notes were changing imperceptibly into original investigations. Within a year, he had digested the achievement of seventeenth-century analysis and had begun to pursue his own independent course into higher analysis.

From the masters he was reading, Newton picked up two of the central problems to which the new analysis, as it was called, addressed itself, drawing tangents to curves (which we have learned to call differentiation) and finding the areas under curves (which they spoke of as quadratures and we know as integration). In Descartes's *Geometry* he found a method of drawing a tangent to a curve at a given point by finding the normal to the curve, which is perpendicular to the tangent, at the point. Quickly Newton mastered the method, noting in a manner typical of him general patterns in analogous equations, and his first success lay in extending Descartes's procedure to finding centers of curvature – or crookedness, in his terminology – and then points of greatest and least curvature. In quadratures he was dependent primarily on the method of infinitesimals as he found it in the works of John Wallis. He was not above making mistakes, but he was also not slow in finding and correcting them as he extended his understanding of the new analysis.

In the winter of 1664–5, or sometime near then, Newton's urge continually to organize his learning led him to draw up a list of "Problems." Initially he put down twelve, one of which he later canceled. He added further problems on several occasions, as different inks show, until he had listed twenty-two in five distinct groups. The first group included most of the problems in analytic geometry to which he had addressed himself so far – to find the axes, diameters, centers, asymptotes, and vertices of lines, to compare their crookedness with that of a circle, to find their greatest and least crookedness, to find the tangents to crooked lines (that is, curves), and so on. The third group looked mostly toward the problems of quadratures to which Wallis had introduced him – to find such lines whose areas, lengths, and centers of gravity may be found, to compare the areas, lengths, and gravities of lines when it can be done, to do the same with the areas, volumes, and gravities of solids, and so on. Several of the problems were mechanical, and one of them treated a curve as the path traced by the end of the line y, perpendicular to x, as the line moves along x. In both respects, the problems looked forward toward distinctive features of his mathematics and of his mechanics. In all, the "Problems" laid out much of the program that would occupy Newton during 1665.

His first important step beyond his mentors, which he dated on several occasions to the winter of 1664–5, was his extension of Wallis's use of infinite series to evaluate areas into what we know as the binomial theorem. Here he drew also upon another new concept, the decimal fraction, which could be used to evaluate a quantity such as pi as closely as one chose by extending the number of places. You should treat quantities calculated by means of binomial expansion into an infinite series, he later explained, "as if you were resolving ye equation in Decimall numbers either by division or extraction of rootes or Vieta's Analyticall resolution of powers; This operation may bee continued at pleasure, ye farther the better. & from each terme ariseing from this operation may bee deduced a parte of ye valor of y." Indeed, delighted with his new discovery, he computed several logarithms from the areas under an equilateral hyperbola to fifty-five places. Adding the binomial theorem, whereby he could express a difficult quantity he wanted to square (or integrate), such as an area equivalent to a logarithm, by an infinite series that he could square term by term, to the established methods of squaring simple powers and polynomials, Newton completed a method by which to find the area under virtually every algebraic curve then known to mathematicians.

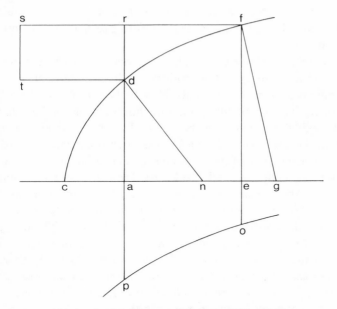

Figure 1. The fundamental theorem of the calculus.

His operations continually displayed patterns. The quadrature of $y =$ x^n is $[1/(n + 1)]x^{n+1}$. Could one not then use this pattern to "shewe y^e nature of another crooked line y^t may be squared?" In the spring of 1665, Newton began seriously to explore the possibilities to which this avenue could lead, bringing the patterns he had observed in determining tangents into contact with the similar but inverted patterns in quadratures. By now surely he had become a stranger to his bed. More than one morning Wickins must have discovered a taut figure bent over his incomprehensible symbols, unaware that a night had passed and for that matter unconcerned. He was repaid by the discovery of the fundamental theorem of the calculus. The problems of tangents and of quadratures suddenly were seen to have an inverse relation to each other. (See Figure 1.)

If the calculus had not been born, certainly it had been conceived. Newton had received his Bachelor of Arts degree, if he had yet received it at all, less than a month before. In mathematics, he had passed further beyond the status of student than a single month can conceivably imply.

He had absorbed by now what books could teach him. Henceforth he would be an independent investigator exploring realms never before seen by human eye.

An essential aspect of the insight was a new approach both to quadratures and to tangents. Before, with Wallis, he had considered areas as static summations of infinitesimals. Now he began to treat them kinetically, as the areas swept out by a moving line. Newton had been uncomfortable with the infinitesimal basis on which his method of tangents rested. In the fall of 1665, he began to extend his kinematic approach to areas to the generation of curves as well and to treat them as the locus of a point moving under defined conditions. From the idea of motion he derived the term *fluxional,* which became his permanent descriptive word for his method. The "infinitely little lines" that bodies describe in each moment are the velocities with which they describe them. The ratio of the velocities of y and x at any point on a curve defines the tangent to that point. The idea of velocity concealed a third, invisible variable, time. The concept of absolute time entered inextricably into Newton's mathematics at this point and found here its permanent rationale in his thought. The concept of continuously varying motion, which appears intuitively to overcome the discontinuity of indivisibles, never ceased to appeal to Newton's imagination. Nevertheless, he would in the end seek another, more rigorous foundation for his calculus.

By now, no doubt, Newton was neglecting his meals. Working in feverish haste, he was ready by 13 November to systematize the new method in a paper entitled "To find ye velocitys of bodys by ye lines they describe." Then, with the completion of the paper, the light went out, as suddenly and as totally as if Newton had extinguished a candle. Six months passed in which, if we can trust the surviving record, he did not turn a finger toward mathematics. In May, something stirred his interest anew, and in two separate papers composed on 14 and 16 May he devoted three days to further elaboration of the idea of motion. Again the light went out, and once more something stirred him in October, when he drew his thoughts together in a more definitive essay. A third time the light went out. It was as though the successful resolution of the problems that had been set for him had exhausted his interest in mathematics. There was no lack of other enthralling enquiries to command his attention. As far as we can tell, he scarcely looked at mathematics for the following two years.

The three papers of 1666 all explore the method based on motion. The second two carry similar titles, finally phrased "To resolve Problems by Motion these following Propositions are sufficient." The latter one of these, known as the tract of October 1666, embodies the definitive statement of Newton's fluxional method, what we know as the infinitesimal calculus.

The tract of October 1666 was a virtuoso performance that would have left the mathematicians of Europe breathless in admiration, envy, and awe. As it happened, only one other mathematician in Europe, Isaac Barrow, even knew that Newton existed, and it is unlikely that in 1666 Barrow had any inkling of his accomplishment. The fact that he was unknown does not alter the other fact that the young man not yet twenty-four, without benefit of formal instruction, had become the leading mathematician of Europe. And the only one who really mattered, Newton himself, understood his position clearly enough. He had studied the acknowledged masters. He knew the limits they could not surpass. He had outstripped them all, and by far.

The tract of October 1666 derived from the insights of 1665. As I understand him, the year 1665 was crucial to Newton's self-awareness. Almost from his first dawning of consciousness, he had experienced his difference from others. Neither in Grantham nor in Cambridge had he been able to mingle successfully with his fellow students. The servants at Woolsthorpe had despised him. Always his insatiable lust to know had set him apart. Now, finally, he had objective proof that his quest for learning was not a delusion. In 1665, as he realized the full extent of his achievement in mathematics, Newton must have felt the burden of genius settle upon him, the terrible burden which he would have to carry in the isolation it imposed for more than sixty years. From this time on, there is little evidence of the futile efforts to ingratiate himself with his peers that appeared intermittently during his grammar school and undergraduate days. Accepting as sufficient his one close relationship with his chamber-fellow Wickins, he abandoned himself, as he had always longed to do, to the imperious demands of Truth.

Though he turned aside from mathematics at the end of 1666, Newton was not by any means done with the method he had created. Significantly, he never tried to publish the tract of October 1666. As he returned to it intermittently in the years ahead, he paid primary attention to improving the foundation of the method; his descriptions of his method at the time of the controversy with Leibniz indicate how far he moved in that respect during about forty years of periodic revision. What he wished to be known for was not what he wrote in 1666, though its inspiration derived directly from the early tract. He also extended the method to recalcitrant problems, such as affected equations, that he could not handle in 1666, and he tackled other areas of mathematics as well. Nevertheless, as he said, he never minded mathematics so intensely again. His great period of mathematical creativity had come to a close. For the most part, his future activities as a mathematician would draw upon the insights of 1665. Years later he told Whiston "that no old Men (excepting Dr. Wallis) love Mathematicks . . . ." True, he was not yet an old man. Other fascinating subjects clamored for the attention of the genius in which he was now confident, however.

✍

Newton was not a man of half-hearted pursuits. When he thought on something, he thought on it continually. By thinking continually on mathematics for a year and a half, he arrived at a new method that allowed him to solve the initial problems, set for him by earlier mathematicians, with which he began. Now other interests represented by the "Quaestiones" could claim his attention. Once they had claimed it, he thought on them as hard as he had on mathematics.

One of these was the science of mechanics. The essay "On violent Motion" in the "Quaestiones" had introduced him to mechanics. There he espoused the doctrine that a force internal to bodies keeps them in motion. In Descartes's *Principles* and in Galileo's *Dialogue,* he confronted the radically different conception of motion that we call today, using language which Newton himself later made common, the principle of inertia. In Descartes he also found two problems posed and imperfectly answered, the mechanics of impact and of circular motion. They became the focus of his investigation.

An early exploration of mechanics, recorded in the *Waste Book,* carried the title "Of Reflections," by which Newton meant impact. A tone of confidence not present in the "Quaestiones" infused the passage. No longer the questioning student, he began to propound alternative solutions. To be sure, he based his treatment of impact squarely on Descartes's conception of motion.

Ax:100 Every thing doth naturally persevere in y^t state in wch it is unlesse it bee interrupted by some externall cause, hence . . . [a] body once moved will always keepe ye same celerity, quantity & determination of its motion.

Of Descartes's law of impact, which completed his discussion of motion, however, Newton did not say a word. He did not even bother to refute it. Instead he launched directly into his own analysis of impact based on a new conception of force. Descartes had analyzed impact in terms of the force internal to a moving body, what he called the "force of a body's motion." In contrast, Newton reasoned that if a body perseveres in its state unless some external cause acts upon it, there must be a rigorous correlation between the external cause and the change it produces. Here was a new approach to force in which a body was treated as the passive subject of external forces impressed upon it instead of the active vehicle of force impinging on others. More than twenty years of patient if intermittent thought would in the end elicit his whole dynamics from this initial insight.

Though all of the possibilities inherent in the insight did not appear immediately to the young man who was then being introduced to the science of mechanics and was grappling with the new conception of motion for the first time, he did succeed in pursuing his insight far enough to realize that any two bodies isolated from external influences constitute a single system whose common center of gravity moves inertially whether or not they impinge on each other. The conclusion is identical to the principle of the conservation of momentum, which remains today the basis for the analysis of impact.

However, complexities associated with the mechanics of circular motion, the second problem posed by Descartes, tended to reinforce his original idea of a force internal to bodies. Following both Descartes and common experience, Newton agreed that a body in circular motion strives constantly to recede from the center, like a stone pulling on its string as it is whirled about. The endeavor to recede appeared to be a tendency internal to a moving body, the manifestation in circular motion of the internal force that keeps a body in motion. Seeking to reduce the tendency to recede to quantitative measure, Newton called upon his recent

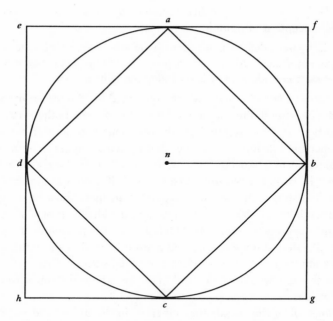

Figure 2. The force of a body moving in a circle derived from impact.

analysis of impact. He imagined that a square circumscribes a circular path and that a body follows a square path inside the circle rebounding at the four points where the circle touches the outer square. (See Figure 2.) From the geometry of the square he was able to compare the force of one impact, in which the component of the body's motion perpendicular to the side it strikes is reversed, to the force of the body's motion, and then to compare the force of the four reflections, the total force in one circuit, to the force of the body's motion. He then generalized the result to polygons of increasing number of sides.

And soe if body were reflected by the sides of an equilaterall circumscribed polygon of an infinite number of sides (i.e. by y^e circle it selfe) y^e force of all y^e reflections are to y^e force of y^e bodys motion as all those sides (i.e., y^e perimeter) to y^e radius.

For one complete revolution the total force F is to the body's motion mv as $2\pi r/r$. Or $F = 2\pi mv$. To see the significance of the result, convert the total force of the body in one revolution to the "force by wch it endeavours from y^e center" at each instant by dividing each side of the equation by the time of one revolution, $2\pi r/u$ The division yields $f = mv^2/r,$ the formula we still use in the mechanics of circular motion.

The formula for a body's endeavor to recede from the center, for which Huygens coined the name "centrifugal force," gave Newton the means to attack a problem that he found in Galileo's *Dialogue*. It was an effort to answer one argument against the Copernican system by showing that the earth's rotation does not fling bodies into the air because the force of gravity, measured by the acceleration of falling bodies, is greater than the centrifugal force arising from the rotation. Newton's solution, chaotically recorded on a piece of parchment, the front side of which had been used by his mother for a lease, was closely associated with the investigations of mechanics in the *Waste Book*. All he needed in addition to his new formula for the force in circular motion were the size of the earth and the acceleration of gravity. For both, he used the figures he found with Galileo's solution of the problem in the Salusbury translation of the *Dialogue*, which appeared in 1665. He arrived at the conclusion "y^t ye force of ye Earth from its center is to ye force of Gravity as one to 144 or there abouts." But why accept Galileo's figure for the acceleration of gravity? He suddenly realized that his measure of centrifugal force opened a further possibility; he could use it to measure g indirectly via a conical pendulum. The measurement, one of the earliest demonstrations of Newton's experimental skill, used the only timekeeper available to him, the sun, together with a conical pendulum 81 inches long inclined at an angle of 45 degrees. It revealed that a body starting from rest falls 200 inches in a second, a figure very close to the one we accept but roughly twice as large as the one he had found in Galileo's *Dialogue*. Hence he returned to his calculation and doubled the ratio of gravity to centrifugal force.

Somewhat later, in a paper which appears to date from the years immediately following his undergraduate career, Newton returned to the same problems. On this occasion, he calculated centrifugal conatus more elegantly by utilizing the geometry of the circle instead of impact. (See Figure 3.) When a body moves in uniform circular motion, time is proportional to length of arc. Because a body will move in a straight line if it is not constrained to move in a circle, Newton set the centrifugal tendency for a brief motion equal to the distance the tangent diverges from the circle. When the arc is "very small," Newton could then apply the known

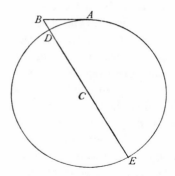

Figure 3. The force of a body moving in a circle derived from its devia-
tion from the tangent.

ratio of the divergence to the arc to calculate the instantaneous force, and
with the force he could calculate the distance it would impel a body in a
straight line, starting from rest, during the time of one revolution. He
employed Galileo's conclusion that distances traversed in uniformly accelerated motion from rest vary as the squares of the times, implicitly
interpreting Galileo's kinematics in dynamic terms. His answer, that in
the time of one revolution the centrifugal force would move a body
through a distance equal to $2\pi^2 r$, is mathematically equivalent to the
earlier formula derived from impact. Again he compared centrifugal force
at the earth's surface to gravity; and because he did not round off his more
accurate measurement of g, he arrived this time at a slightly higher ratio,
1:350.

So far he had come before. He was ready now to take a further step. He
compared the "endeavour of the Moon to recede from the centre of the
Earth" with the force of gravity at the surface of the earth. He found that
gravity is somewhat more than 4,000 times as great. He also substituted
Kepler's third law (that the cubes of the mean radii of the planets vary as
the squares of their periods) in his formula for centrifugal force: "the
endeavours of receding from the Sun [he discovered] will be reciprocally
as the squares of the distances from the Sun." Here was the inverse-
square relation resting squarely on Kepler's third law and the mechanics
of circular motion. To catch the full significance of the statement one

must reflect on the earlier ratio of gravity to the moon's tendency to recede from the earth. He had found a ratio of about 4,000:1. Because he was using 60 earth radii as the moon's distance, the exact ratio according to the inverse-square relation should have been 3,600:1. It is difficult to believe that this paper was not what Newton referred to when he said that he found the comparison of the force holding the moon in its orbit to gravity to "answer pretty nearly."

What then is one to make of the story of the apple? It is too well attested to be thrown out of court. In Conduitt's version, one of four independent ones, it runs as follows:

In the year 1666 he retired again from Cambridge . . . to his mother in Lincolnshire & whilst he was musing in a garden it came into his thought that the power of gravity (wch brought an apple from the tree to the ground) was not limited to a certain distance from the earth but that this power must extend much farther than was usually thought. Why not as high as the moon said he to himself & if so that must influence her motion & perhaps retain her in her orbit, whereupon he fell a calculating what would be the effect of that supposition but being absent from books & taking the common estimate in use among Geographers & our seamen before Norwood had measured the earth, that 60 English miles were contained in one degree of latitude on the surface of the Earth his computation did not agree with his theory & inclined him then to entertain a notion that together with the force of gravity there might be a mixture of that force wch the moon would have if it was carried along in a vortex . . . .

Small wonder that such an anecdote, redolent of the Judaeo-Christian association of the apple with knowledge, continues to be repeated. Together with the myth of the *annus mirabilis* and with Newton's memorandum that said he found the calculation to answer pretty nearly, it has contributed to the notion that universal gravitation appeared to Newton in a flash of insight in 1666 and that he carried the *Principia* about with him essentially complete for twenty years until Halley pried it loose and gave it to the world. Put in this form, the story does not survive comparison with the record of his early work in mechanics. The story vulgarizes universal gravitation by treating it as a bright idea. A bright idea cannot shape a scientific tradition. Lagrange did not call Newton the most fortunate man in history because he had a flash of insight. Universal gravitation did not yield to Newton at his first effort. He hesitated and floundered, baffled for the moment by overwhelming complexities, which were great enough in mechanics alone and were multiplied sevenfold by the total context. What after all was in the paper that revealed the inverse-square relation? Certainly not the idea of universal gravitation. The paper spoke only of tendencies to recede, and to Newton the mechanical philosopher an attraction at a distance was inadmissible in any case. Revealingly, Conduitt brought in the vortex. Nevertheless, Newton must have had something in mind when he compared the moon's centrifugal force with gravity, and there is every reason to believe that the fall of an apple gave rise to it. Though he did not name the force explicitly, something had to press back on the moon if it remained in orbit. Something had to press back on the planets. Moreover, Newton remembered both the occasion and the calculation, so that fifty and more years later they seemed to constitute an important event in his development. Some idea floated at the border of his consciousness, not yet fully formulated, not perfectly focused, but solid enough not to disappear. He was a young man. He had time to think on it as matters of great moment require.

❧

Motion and mechanics were not the only topics in natural philosophy that commanded Newton's interest. As important in his own eyes were what he later called the "celebrated Phaenomena of Colours." The phenomena of colors had become a celebrated topic in optics for at least two reasons. What we call chromatic aberration appeared in every telescopic observation, coloring the images and confusing their focus. In contrast, colors sharply focused the different stances of the Aristotelian and the mechanical philosophies of nature. It is scarcely surprising that colors were among the "Quaestiones quaedam Philosophicae" compiled by the young mechanical philosopher in Cambridge. He had found the issue in Descartes, in Boyle's *Experiments and Considerations Touching Colors* (1664), and in Hooke's *Micrographia* (1665). Dissatisfied with their explanations of colors, as his notes show, he turned his hand to his own.

After the passage in the "Quaestiones" Newton returned to colors again, probably in 1665, utilizing empty pages at the end of his original set of headings. There is a good likelihood that the theory of colors in Robert Hooke's *Micrographia* stimulated him. His immediate negative reaction to Hooke's account inaugurated forty years of antipathy between two incompatible men. As in mechanics, he was not content any longer simply to question. An alternative theory sprang to mind. Hooke proposed that "Blue is an impression on the Retina of an oblique and confus'd pulse of light, whose weakest part precedes, and whose strongest follows." Red is the impression of "an oblique and confus'd pulse" of reversed order. On the first page of his new set of notes, Newton contradicted the two fundamental assertions of Hooke's theory, that light consists of pulses and that colors arise from confused impressions. "The more uniformly the globuli move ye optick nerves ye more bodys seme to be coloured red yellow blue greene &c but ye more variously they move them the more bodys appear white black or Greys." If Hooke was the immediate target of the assertion, much more than Hooke's theory was involved. Like other mechanical philosophers, Hooke had merely provided a mechanism for the existing theory of apparent colors, phenomena such as the rainbow and the colored fringes seen through telescopes and prisms. The theory had been fatally easy to mechanize. It had employed a scale of colors, which was also a scale of strength, running from brilliant red, considered to be pure white light with the least admixture of darkness, to dull blue, the last step before black, which was the complete extinction of light by darkness.

The proposal of the freshly minted Bachelor of Arts implied a completely different relation of light and color. White light, ordinary sunlight, is a confused mixture. Individual components of the mixture, which he considered to be corpuscles rather than pulses, cause sensations of individual colors when they are separated from the mixture and fall on the retina alone. Already he had drawn a picture of an eye looking through a prism at the colored fringes along a border between black and white. From the two sides of the border two rays were shown following different paths through the prism as they were refracted at different angles and emerged along the same line incident on the eye. "Note y^t slowly moved rays are refracted more then swift ones." Though he would modify details as he clarified his understanding of its implications, the comment contains the insight on which Newton built his work in optics. The insight fundamental to his dynamics had happened less than a year before, both of them less than two years after he turned seriously to natural philosophy. He had a keen eye for the critical point at which to seize a problem.

He started with an idea rather than an observation. Under the diagram of the prism and the eye was a table in which he tried to reason out the

colors that would appear along the borders between various combinations
other than black and white. Quickly the complexities of mentally sorting
out slowly and swiftly moving rays reflected from various bands along the
border became more confused than Hooke's pulses. As suddenly as the
original insight, a simplifying experiment presented itself. An experimental orientation pervaded the "Quaestiones," but until the investigation of
colors experimentation had been implicit rather than explicit, questions
put but not experiments performed. At this point the period of adolescence was fulfilled, and Newton the experimental scientist reached matu-
rity.

That ye rays wch make blew are refracted more y^n ye rays wch make red appeares
from this experim^{nt}. If one hafe [one end] of ye thred *abc* be blew & ye other red &
a shade or black body be put behind it y^n lookeing on ye thred through a prism one
halfe of ye thred shall appear higher y^n ye other. & not both in one direct line, by
reason of unequall refractions in ye 2 differing colours.

The idea had been provisionally confirmed. Newton never forgot this
experiment; he continued to cite it as one of the basic supports of his
theory of color.

At the time he made the experiment, however, the theory hardly existed. It was only a promising idea supported by a single experiment. Its
implications are obvious to us who profit from three hundred years of
digesting them. Newton had to grope his way forward as he denied a
tradition two thousand years old which seemed to embody the dictates of
common sense. The concept of slow and swift rays was formulated within
the context of a mechanical philosophy, and it carried the usual connota-
tions of weak and strong. It inclined him to think in terms of a two-color
system, blue and red. It inclined him as well to imagine mechanisms by
which the "elastick power" of a body's particles determined how much of
the motion of a ray was reflected; "then y^t body may be lighter or darker
colored according as ye elastick virtue of that bodys parts is more or
lesse." Such ideas returned to the assumption that colors arise from the
modification of light, against which his central insight was directed.

Perhaps it was here that his consideration of devices to grind elliptical
and hyperbolic lenses, which he mentioned in his paper of 1672, intervened. The background of the investigation was Descartes's announcement of the sine law of refraction in his *Dioptrique*. As use of the telescope
had spread in the early seventeenth century, experience had shown that

spherical lenses do not refract parallel rays, such as those from celestial bodies, to a perfect focus. In *La Dioptrique,* Descartes had shown that hyperbolic and elliptical lenses would do so, given the sine law of refraction. Grinding them was another question. Spherical surfaces present no problem. Because they are symmetrical in all directions, constant turning and shifting of a lens adjust the lens and the form against which it is being ground to each other so that a spherical surface is bound to result. On the other hand, the grinding of an elliptical or hyperbolic surface is complicated indeed, exactly the problem to challenge the model-builder from Grantham, who was now equipped with a thorough knowledge of the conics. He sketched out several devices by which to produce them. And as he did so, he possibly reflected on the meaning of his earlier experiment with the prism and the red and blue thread. Descartes's demonstration had assumed the homogeneity of light. What if Newton did succeed in grinding elliptical and hyperbolic lenses? He still would not obtain a perfect focus because light is not homogeneous; the blue rays are refracted more than the red. At this moment, it appears to me, Newton began to realize the significance of his experiment and of the idea behind it. He stopped working on nonspherical lenses and never returned to them. Later he showed that chromatic aberration introduces much larger errors in lenses than does spherical aberration. Instead of lenses, he turned his attention to an experimental investigation of the heterogeneity of light and its role in the production of colors. I assume that this investigation came in 1666 and was what he referred to later. Only with it did he "have the Theory of Colours" in any legitimate sense of the phrase.

Newton recorded the investigation as the essay "Of Colours" in a new notebook in which he extended several of the topics of the "Quaestiones." With his purpose now more clearly focused, he marshaled known phenomena of colors, which he had found in Boyle and Hooke, that exhibited the analysis of light into its components. Thus thin leaves of gold appear yellow from one side in reflected light but blue in transmitted light from the other; with a solution of *lignum nephriticum* (nephritic wood, infusions of which were used medicinally at the time) the colors are the reverse. In both cases, the transmission of some rays and reflection of others analyzes white light into its components. Newton was convinced that all solid bodies would behave like gold if pieces thin enough could be obtained, and that the solution of *lignum nephriticum* would appear blue from all sides if it were made thick enough so that no light could pass through. If Newton turned available observations to the advantage of his theory, however, he relied primarily on his own experiments with the prism. Guided by his ingenuity, the prism became an instrument of precision with which he dissected light into its elementary components. No other investigation of the seventeenth century better reveals the power of experimental enquiry animated by a powerful imagination and controlled by rigorous logic.

Among the authors Newton had read, both Boyle and Hooke had employed variations of Descartes's projection of a prismatic spectrum to examine colors. Newton saw that he could bend the same experiment to test his own theory by imposing carefully prescribed conditions on it. If in fact light is heterogeneous and different rays are refracted at different angles, a round beam should be projected by a prism into an elongated spectrum. It would require enough distance to spread out, however. Rays are ideal entities; in actual experimentation a physical beam had to be used, and one big enough to give visible effects. If the screen were placed close to the prism as it had been in earlier experiments, the expected elongation would not appear. Descartes had received his spectrum on a screen only a few inches from the prism. Hooke, who employed a deep beaker filled with water instead of a prism, had about two feet between refraction and screen. Boyle apparently used the floor and hence had a distance of perhaps four feet. Newton projected his spectrum onto a wall twenty-two feet away. Where earlier investigators saw a spot of light colored at its two edges, Newton saw a spectrum five times as long as it was wide. His exposition in 1672 contrived to suggest an element of chance and surprise when he saw it; it was as accidental as the observation of a barometer by Pascal's brother-in-law on the summit of the Puy de Dôme. Newton had constructed his experiment to test what he wanted to test. Had the spectrum not been elongated, his promising idea would have been refuted at its second step, and he could not have elaborated it into a theory.

If he had not refuted himself, he was far from having proved anything, as he knew very well. What he proposed was a radical reordering of the relation of light and color. Whereas the received opinion considered white light as simple and colors as modifications of it, Newton asserted that the light which provokes the sensations of individual colors is simple and white light a complex mixture. Long-established views are not easily surrendered. Possible objections were many; they would need to be answered. By testing, he showed, for example, that irregularities in the glass could not have caused the elongation.

The most important objection of all was mathematical. Because the sun fills a visual angle of 31 minutes, the beam incident on the prism was not composed of parallel rays. By the sine law of refraction, rays incident at different angles are refracted at different angles. Could the elongated spectrum be an unexpected product of the sine law? Newton employed various devices to get a beam composed of rays more nearly parallel, but he knew that only a theoretical demonstration could finally meet the objection. It was not a difficult exercise for a mathematician of his accomplishment. When the central ray of an incident pencil of homogeneous light contained within an angle of 31 minutes is refracted equally at both faces of a prism, it emerges as a pencil contained within an angle of 31 minutes. As it happens, equal refraction at each face is also the condition for minimal refraction, so that Newton had only to turn the prism until the spectrum reached its lowest position on the wall to obtain it. His first recorded projection of a spectrum noted that the rays were equally refracted by both faces of the prism. Along with the distance of projection, equal refraction at the two faces was a planned condition of the initial experiment. So far was the elongated spectrum from a chance observation.

Although the mathematical demonstration supplied necessary rigor to the evidence of the spectrum, Newton ultimately found a further experiment that seemed to confirm it no less strongly. Like the final theory, the experiment did not appear in one burst of inspiration. It evolved through several stages until, as he realized its force, he called it the *experimentum crucis*. In its initial form, it was ill defined and hardly compelling. He simply held a second prism in the spreading spectrum five or six yards from the first. The blue rays suffered a greater refraction than the red. In neither case did the second refraction produce further coloration; blue remained blue, and red remained red. In 1666 he went no further. Only later did he realize the demonstrative potential he could gain by refining the experiment.

Once Newton had fully seized the concept of analysis, he was able readily to generate other experiments to illustrate it. He could analyze sunlight into its components by tilting a prism to the critical angle, where blue rays, which are the most refrangible, began to be reflected from the second face while red rays were still transmitted through it. He obtained an analogous separation with a thin film of air trapped between two prisms bound together. Realizing that it was necessary to demonstrate that he could reconstitute white, he cast the spectra of three prisms onto one another so that they overlapped without coinciding. In the center, where all the colors fell, the combined spectrum was white. He attached a paper to the face of a prism with several slits parallel to the edges. On a screen held near the prism a line of color appeared for each slit. As he moved the screen away, the center of the spectrum became white, but the full spectrum appeared again, without further experimental manipulation, as he moved the screen still farther.

Some seven years later, after the publication of his first paper in the *Philosophical Transactions,* Newton replied to a critique by the Dutch scientist Christiaan Huygens with a methodological homily.

It seems to me that M. Hugens takes an improper way of examining the nature of colours whilst he proceeds upon compounding those that are already compounded, as he doth in the former part of his letter. Perhaps he would sooner satisfy himself by resolving light into colours as far as may be done by Art, and then by examining the properties of those colours apart, and afterwards by trying the effects of reconjoyning two or more or all of those, & lastly by separating them again to examin w^t changes that reconjunction had wrought in them. This will prove a tedious & difficult task to do it as it ought to be done but I could not be satisfied till I had gone through it.

No doubt Huygens, the doyen of European science, did not relish the lecture from an unknown professor in Cambridge. Nevertheless it is a reasonable description of Newton's procedure as he unraveled the implications of his central idea.

One other issue remained: the colors of solid bodies. Newton built his theory of colors from experiments with prisms. The vast majority of the colors we see, however, are associated with solid bodies. Unless he accounted for their colors, his theory would be extremely limited. From the moment of his initial insight, of course, he had a general account of the colors of solid bodies. Reflection can also analyze white light into its components. A body is disposed to reflect some rays more than others and appears to be the color it reflects best. His account of colors never deviated from this position. In the beginning, however, the statement expressed an idea without empirical foundation and without quantitative content. The essay "Of Colours" provided some empirical foundation. When he painted red and blue patches on a piece of paper and viewed them in "Prismaticall blew" and "Prismaticall red," both patches appeared to have the color of the incident light, but the blue patch was fainter in red light and the red fainter in blue. "Note y^t ye purer ye Red/Blew is ye lesse tis visible wth blew/Red rays." Later he would add further empirical evidence as the immutability of rays became clearer to him.

Quantitative content was a more difficult matter. It was absolutely essential. After his experiments with prismatic spectra, analysis by refraction could be expressed in rigorous quantitative terms. Color ceased to be a wholly subjective phenomenon because it was attached immutably to a given degree of refrangibility. With reflected colors, in contrast, he had attained no similar quantitative treatment, and reflected colors constitute the overwhelming bulk of color phenomena in the world. He had picked up a suggestion, however. In Hooke's *Micrographia* he found descriptions of colors in a variety of thin transparent bodies – in Muscovy glass (or mica), in soap bubbles, in the scoria of metals, in the air between two pieces of glass. Newton himself observed the colors in a film of air between two prisms, both in the transmitted and in the reflected light. The "plate of air (*ef*) is a very reflecting body," he noted, and later he indicated that the colors of solid bodies are related to the colors of thin transparent films. He even realized a means of doing what Hooke had confessed himself unable to do, measuring the thickness of the films in which colors appear. When a lens of known curvature was pressed on a flat piece of glass, a thin film of air was constituted between them. Circles of color appeared around the point of contact. Using the geometry of the circle – the same proposition, indeed, that he used in calculating centrifugal force – he computed the thickness of the film from the curvature of the lens and the measured diameter of the circles. "Of Colours" recorded Newton's first observation of "Newton's rings."

❧

On close examination, the *anni mirabiles* turn out to be less miraculous than the *annus mirabilis* of Newtonian myth. When 1666 closed Newton was not in command of the results that have made his reputation deathless, not in mathematics, not in mechanics, not in optics. What he had done in all three was to lay foundations, some more extensive than others, on which he could build with assurance, but nothing was complete at the end of 1666, and most were not even close to complete. Far from diminishing Newton's stature, such a judgment enhances it by treating his achievement as a human drama of toil and struggle rather than as a tale of divine revelation. "I keep the subject constantly before me," he said, "and wait 'till the first dawnings open slowly, by little and little, into a full and clear light." In 1666, by dint of keeping subjects constantly before him, he saw the first dawnings open slowly. Years of thinking on them continuously had yet to pass before he gazed on a full and clear light.

By any other standard than Newtonian myth, the accomplishment of the *anni mirabiles* was astonishing. In 1660, a provincial boy ate his heart out for the world of learning which he was apparently being denied. By good fortune it had been spread before him. Six years later, with no help beyond the books he had found for himself, he had made himself the foremost mathematician in Europe and the equal of the foremost natural philosopher. What is equally important for Newton, he recognized his own capacity because he understood the significance of his achievements. He did not merely measure himself against the standard of Restoration Cambridge; he measured himself against the leaders of European science whose books he read. In full confidence he could tell the Royal Society early in 1672 that he had made "the oddest if not the most considerable detection wch hath hitherto beene made in the operations of Nature."

The parallel between Newton and Huygens in natural philosophy is remarkable. Working within the same tradition, they saw the same problems in many cases and pursued them to similar conclusions. Beyond mechanics, there were also parallel investigations in optics. At nearly the same time and stimulated by the same book, Hooke's *Micrographia,* they thought of identical methods to measure the thickness of thin colored films. No other natural philosopher even approached their level. In the very year 1666, Huygens with all his acclaim was being wooed by Louis XIV to confirm the renown of his Académie Royale des Sciences. There was no occasion for a young man recently elevated to the dignity of Bachelor of Arts and working in isolation to be ashamed of his achievement, even if the Sun King, in his presumption, had not placed a crown of laurel on his brow.

4

Lucasian Professor

SHORTLY AFTER HIS RETURN from Woolsthorpe late in April 1667, the magnificent funeral of Matthew Wren, bishop of Ely, escorted by the entire academic community in full regalia according to their ranks and degrees, must have reminded Newton that he stood then only on the first step of the university hierarchy and that others loomed immediately ahead. In only a few months he would face the first and by far the most important of these, the fellowship election. As with the scholarship three years earlier, Newton's whole future hung in the balance of this election. It would determine whether he would stay on at Cambridge and be free to pursue his studies or whether he would return to Lincolnshire, probably to the village vicarage that his family connections could have supplied, where he might well have withered and decayed in the absence of books and the distraction of petty obligations. On the face of it, his chances were slim. There had been no elections in Trinity for three years, and, as it turned out, there were only nine places to fill. The phalanx of Westminster scholars exercised their usual advantage. The growing role of political influence, whereby those with access to the court won letters mandate from the king commanding their election, was notorious. For the rest, all depended on the choice of the master and eight senior follows, and stories of influence peddling filled the air. The candidates had to sit in the chapel four days in the last week of September to be examined *viva voce* by the senior fellows, the dying embodiment of the curriculum Newton had systematically ignored for nearly four years. How could an erstwhile subsizar of whatever capacity hope to prevail against such odds? If he too had a patron, he might do more than hope. In 1667, Humphrey Babington joined the ranks of the senior fellows. Neither in Newton's papers nor in the surviving anecdotes does a hint of tension over the outcome appear. He spent £1 10s on tools, real tools, including a lathe, such as he must have longed for in Grantham – not the purchase of a man seriously expecting to move on a year hence. And Newton invested handsomely in cloth for a bachelor's gown, which could be converted later into a master's, eight-and-a-half yards of "Woosted Prunella" plus four yards of lining, for which he paid nearly £2 in all. On 1 October a bell tolled at eight in the morning to summon the seniors to the election. The bell tolled again the following day at one to call those chosen to be sworn in: It tolled for Newton.

Now at last the way was clear. The election promised permanent membership in the academic community with freedom to continue the studies so auspiciously, as he at least understood, begun. True, two more steps remained for him to mount. In October 1667, he became only a minor fellow of the college, but advancement to the status of major fellow would follow automatically when he was created Master of Arts nine months hence. The exercises for the degree had become wholly *pro forma;* no one was known to have been rejected. The final step could come at any time in the following seven years. The incumbents of two specific fellowships excepted, the sixty fellows of the college were required to take holy orders in the Anglican church within seven years of incepting M.A. Shortly after one o'clock on 2 October 1667, Newton became a fellow of the College of the Holy and Undivided Trinity when he swore "that I will embrace the true religion of Christ with all my soul . . . and also that I will either set Theology as the object of my studies and will take holy orders when the time prescribed by these statutes arrives, or I will resign from the college." The final requirement was not likely to pose more of an obstacle to a pious and earnest young man than the Master's degree.

After his creation as Master of Arts, Newton lived in Trinity for twenty-eight years. Those years coincided roughly with the most disastrous period in the history both of the college and of the university. Whatever his initial expectations may have been, he did not find a congenial circle of fellow scholars. A philosopher in search of truth, he found himself among placemen in search of a place. This fundamental fact colored the scene in which virtually the whole of his creative life was set.

Against this background, we can read the various anecdotes that have survived about his life in the college. Many of them derived from

Humphrey Newton (no relation), who served as Newton's amanuensis in Cambridge for five years in the 1680s. This was a unique period in Newton's life, while he was composing the *Principia*. Perhaps we should exercise some caution in treating Humphrey's recollections as typical, though Newton's capacity to be dominated by a problem did not confine itself to the *Principia*. Stories of Newton's absent-mindedness were rife in Cambridge, as William Stukeley, a student at Cambridge early in the eighteenth century and later a friend of Newton, also reported.

As when he has been in the hall at dinner, he has quite neglected to help himself, and the cloth has been taken away before he has eaten anything. That sometime, when on surplice days, he would goe toward S. Mary's church, instead of college chapel, or perhaps has gone in his surplice to dinner in the hall. That when he had friends to entertain at his chamber, if he stept in to his study for a bottle of wine, and a thought came into his head, he would sit down to paper and forget his friends.

Humphrey Newton's chaotic stream of consciousness contained similar recollections.

He alwayes kept close to his studyes, very rarely went a visiting, & had as few Visiters, excepting 2 or 3 Persons, Mr Ellis of Keys, Mr Lougham [called Laughton in his other letter] of Trinity, & Mr Vigani, a Chymist, in whose Company he took much Delight and Pleasure at an Evening, when he came to wait upon Him. I never knew him take any Recreation or Pastime, either in Riding out to take ye Air, Walking, Bowling, or any other Exercise whatever, Thinking all Hours lost, yt was not spent in his studyes, to wch he kept so close, yt he seldom left his Chamber, unless at Term Time, when he read in ye schools, as being Lucasianus Professor . . . . He very rarely went to Dine in ye Hall unless upon some Publick Dayes, & then, if He has not been minded, would go very carelesly, wth Shooes down at Heels, Stockins unty'd, surplice on, & his Head scarcely comb'd.

"He would with great acutness answer a Question," Humphrey added in his second letter, "but would very seldom start one." During five years, Humphrey saw Newton laugh only once. He had loaned an acquaintance a copy of Euclid. The acquaintance asked what use its study would be to him. "Upon which Sir Isaac was very merry."

It is not hard to recognize in these anecdotes the man who unconsciously sketched his own portrait in his papers, a man ravished by the desire to know. Equally, it is not hard to recognize his status in Trinity – isolation, indeed alienation. True, Stukeley mentioned friends being entertained in his chamber, and Humphrey Newton named three of them. The references scarcely suffice to erase the impression left by the rest. Newton seldom leaves his chamber. He prefers to eat there alone. When he does dine in the hall, he is hardly a genial companion; rather, he sits silently, never initiating a conversation, as isolated in his private world as though he had not come. He does not join the fellows on the bowling green. He rarely visits others. None of those who visit him are fellows of Trinity. Of the three, we know that Newton later broke with Vigani because he "told a loose story about a Nun . . . ." Whatever his friendships with Laughton and Ellis, they were not close enough to elicit correspondence from either side after Newton left Cambridge.

On 18 May 1669, Newton wrote a letter to Francis Aston, a fellow of Trinity who had received leave to travel abroad and was then departing. According to the letter, Aston had asked his advice about traveling, and Newton had complied rather fully. The bulk of the letter was worldly advice cribbed from a discourse still found among the Newton papers: "An Abridgement of a Manuscript of Sr Robert Southwell's concerning travelling." Aston should adapt his behavior to the company he is in. He should ask questions but not dispute. He should praise what he sees rather than criticize. He should realize that it is dangerous to take offense too readily abroad. He should observe various things about the economy, society, and government of the countries he visits. A final paragraph added a number of particular enquiries that Newton wished Aston to make, mostly about alchemy and mostly based on Michael Maier, *Symbola aureae mensae duodecim nationum* (Frankfort, 1617). The letter to Aston is among the most eloquent in Newton's correspondence. Not for its content – with its borrowed air of worldliness, the letter itself is more ludicrous than eloquent. It is found today among Newton's own papers, which suggests that he recognized he was cutting a ridiculous figure as he assumed a worldly posture on the basis of one month in London and an essay by Southwell, and decided not to send it. The eloquence of the letter lies in its uniqueness. It is the only personal letter to or from a peer in Cambridge in the whole corpus of Newton's correspondence. In its uniqueness, it adds color to the portrait of isolation in Stukeley's and Humphrey Newton's anecdotes.

So do surviving accounts of the college that were collected in the second decade of the eighteenth century. Thomas Parne, B.A. 1718, collected materials for a history of the college, including the recollections of elderly fellows such as George Modd. He recorded particulars about Ray, Pearson, Barrow, Thorndike, and Duport. Newton was a famous man when Parne was drawing his materials together, far more famous than the men just mentioned, but only three references to him appear in the collection: his name (without any comment whatever) at the head of the list of writers, the dates of his elections to Parliament and of his later unsuccessful bid for election, and one brief anecdote about his absence of mind. At much the same time, James Paine, who was elected to a fellowship in 1721, set down a conversation with Robert Creighton, who had been a fellow from 1659 to 1672. Creighton recalled Pearson, Dryden, Gale, Wilkins, and Barrow; he did not mention Newton. Neither did Samuel Newton (no relation), who as registrar and auditor was employed by the college during the whole of Newton's tenure as fellow, enter his name in the diary he kept until Newton's election to Parliament in 1689. To be sure, Samuel Newton filled his diary more with events than with references to individuals. Nevertheless, it seems evident that Newton did not loom prominently on the college scene.

Two stories do suggest that for their part the other fellows, whatever their amusement at his absent-mindedness, regarded him with awe. In 1667, he seemed to have prophetic powers when the Dutch fleet invaded the Thames.

Their guns were heard as far as Cambridg, and the cause was well known; but the event was only cognisable to Sir Isaac's sagacity, who boldly pronounc'd that they had beaten us. The news soon confirm'd it, and the curious would not be easy whilst Sir Isaac satisfy'd them of the mode of his intelligence, which was this; by carefully attending to the sound, he found it grew louder and louder, consequently came nearer; from whence he rightly infer'd that the Dutch were victors.

When he walked in the fellow's garden, "if some new gravel happen'd to be laid on the walks, it was sure to be drawn over and over with a bit of stick, in Sir Isaac's diagrams; which the Fellows would cautiously spare by walking beside them, and there they would sometime remain for a good while."

As far as we know, Newton formed only three close connections in Trinity, all of them rather elusive. With John Wickins, the young pensioner he met on a solitary walk in the college, he continued to share a chamber until Wickins resigned his fellowship in 1683 for the vicarage of

Stoke Edith. Apparently Newton severed effective communication with Wickins when he left Trinity, and we know curiously little about Newton's closest friendship during twenty critical years of his life. In addition to Wickins, there were relationships with Humphrey Babington and Isaac Barrow about which we know even less.

The alienation from college society worked to Newton's advantage. The increasing triviality of the fellows' lives could entangle a promising man and destroy him. Passionately inclined to study in any case, Newton turned away from his peers and in upon himself and surrendered completely to the pursuit of knowledge. The Steward's Books of the college show that he seldom left. In 1669 (which means, in the college books, the twelve months that ended with Michaelmas, 29 September 1669) he was there all fifty-two weeks; in 1670, forty-nine-and-a-half; in 1671, forty-eight; in 1672, forty-eight-and-a-half. When he did leave, it was usually for a trip home. A decade later, Humphrey Newton found that he seldom went to morning chapel because he studied until two or three every morning. For that matter, he seldom interrupted his studies to attend evening chapel either, though he did go to church in St. Mary's on Sundays. "I believe he grudg'd y^t short Time he spent in eating & sleeping," Humphrey Newton observed. The Reverend John North, master of the college from 1677 to 1683 and resident in it for a time before that, who rather fancied himself a scholar, "believed if Sir Isaac Newton had not wrought with his hands in making experiments, he had killed himself with study."

The laxity of the system, which had helped him already as an undergraduate, continued to favor him. If it demanded nothing of fellows such as George Modd and Patrick Cock, Newton's contemporaries who vegetated in the college for forty years without either teaching or pursuing research, likewise it demanded nothing of him. His use of his leisure may have unsettled the others, but the essence of the system was tolerance. Unrelenting study on a fellowship intended to support study was not demonstrably more subversive than drawing dividends in absentia. Supported comfortably, Newton was free to devote himself wholly to whatever he chose. To remain on, he had only to avoid the three unforgivable sins: crime, heresy, and marriage. Safely ensconced with Wickins in the orthodox fastness of Trinity College, he was not likely to sacrifice his security for one of them.

ॐ

Besides the three topics of the *anni mirabiles,* a new subject began now to engross him. His accounts show that in 1669 he spent 14s for "Glasses" in Cambridge and 15s more for the same in London. He made some other purchases in London as well.

For Aqua Fortis, sublimate, oyle perle [*sic* – per se?] fine Silver, Antimony, vinegar Spirit of Wine, White lead, Allome Nitre, Salt of Tartar, ☿ [mercury]	2.	0.	0
A Furnace	0.	8.	0
A tin Furnace	0.	7.	0
Joyner	0.	6.	0
Theatrum Chemicum	1.	8.	0

He also paid 2s to have the oil transported to Cambridge. "Theatrum Chemicum" referred to the huge compilation of alchemical treatises published by Lazarus Zetzner in 1602 and recently expanded to six volumes. More than woodworking was going on in the chamber shared by the long-suffering Wickins. Years later Newton remarked to Conduitt that Wickins, who was stronger than he, used to help him with his kettle, "for he had several furnaces in his own chambers for chymical experiments." When Newton's hair turned gray early in the 1670s, Wickins told him it was the effect of his concentration. Newton, whom Humphrey Newton saw laugh only once, would jest that it was "ye Experim^{ts} he made so often wth Quick Silver, as if from Hence he took so soon that Colour."

Meanwhile, chemistry was not his only study. In 1669, events focused Newton's attention once more on his fluxional method and forced him to take it from his desk. Though he did not then publish it, at least he made it known. Toward the end of 1668, Nicholas Mercator published a book, *Logarithmotechnia,* in which he gave the series for log (1 + x), which he had derived by simply dividing 1 by (1 + x) and squaring the series term by term. As the title suggests, he realized that the series offered a simplified means to calculate logarithms. Some months later – the exact time is unknown, but it appears from the dates of following events to have been in the early months of 1669 – John Collins sent a copy of the book to Isaac Barrow in Cambridge. Collins was a mathematical impresario who had made it his business to foster his favorite study. To that end, he functioned as a clearinghouse for information, attempting by his correspondence to keep the growing mathematical community of England and Europe abreast of the latest developments. No doubt Collins was playing this role when he sent a copy of Mercator's work to the Lucasian professor of mathematics. Late in July, Collins received in reply a letter which informed him that a friend of Barrow's in Cambridge, "that hath a very excellent genius to those things, brought me the other day some papers, wherein he hath sett downe methods of calculating the dimensions of magnitudes like that of Mr Mercator concerning the hyperbola, but very generall . . . ." Barrow was not mistaken in thinking the paper would please Collins, and he promised to send it with his next letter. About ten days later, Collins did receive a paper with the title *De analysi per aequationes numero terminorum infinitas (On Analysis by Infinite Series).* Late in August, he learned who the author was. "His name is Mr Newton; a fellow of our College, & very young (being but the second yeest Master of Arts) but of an extraordinary genius & proficiency in these things."

Among other things, the episode informs us that Barrow and Newton were by this time acquainted. Apparently they had been for some time; Collins would later note that Newton contrived a general method of infinite series "above two yeares before Mercator published any thing, and communicated the same to Dr Barrow, who accordingly hath attested the same." Hence when Barrow received Mercator's book, he realized its implication for Newton's work and showed it to him.

The episode served also to confront Newton with the enormous anxieties that prospective publication aroused. Describing the events a few years later, he said that upon the appearance of Mercator's book, "I began to pay less attention to these things, suspecting that either he knew the extraction of roots as well as division of fractions, or at least that others upon the discovery of division would find out the rest [of the binomial expansion] before I could reach a ripe age for writing." Forget the main and the final clauses; they imposed later reflections on his initial reaction. What he found in Mercator's book was half of the discovery that had set him on his way four years earlier. If Mercator had done it for the hyperbola, would he not do it for the circle as well (i.e., the series for $(1 - x^2)^{1/2}$, "the extraction of roots")? Moreover, Mercator had applied series expansion to quadratures. To Newton at least, the whole of his proud advance stretched out directly beyond the door Mercator had opened. We know from Collins's correspondence that others caught the published hint. Lord Brouncker claimed to have found a series for the area of a circle. James

Gregory was working toward one. More than once, Mercator himself claimed to have one. It is unlikely that Newton had heard of these claims, but his imagination would have filled them in, for he knew that infinite series were in the air and that other mathematicians were at work. In haste, he composed a treatise, drawn from his earlier papers, which by its generality (in contrast to Mercator's single series) would assert his priority. Still in haste, he took it to Barrow, who proposed to do the obvious thing and send it to Collins. As he faced the implications of that move, Newton's haste melted suddenly away. Hitherto he had communed with himself alone, aware of his own achievement but secure from profane criticism. Lately he had communicated something to Barrow, but Barrow was a member of the closed society of Trinity, the only one in it able even to read his paper. What now opened before Newton was something much more, and apparently he shrank back in fright, all too aware of his unripe age and much else besides. When Barrow wrote on 20 July, he had the paper in his possession but was not allowed to send it. We can only imagine what went on during the following days, though his letter of 31 July gives more than a hint.

I send you the papers of my friend I promised . . . . I pray having perused them as much as you thinke good, remand them to me; according to his desire, when I asked him the liberty to impart them to you. and I pray give me notice of your receiving them with your soonest convenience; that I may be satisfyed of their reception; because I am afraid of them; venturing them by the post, that I may not longer delay to corrispond with your desire.

Only when Collins's enthusiastic response calmed his fears did Newton allow Barrow to divulge his name and give permission for Brouncker to see the paper. In all, it was a fair show of apprehension by a man who knew himself to be the leading mathematician of Europe.

As both its title and the circumstances of its composition suggest, *De analysi* concerned itself primarily with infinite series in their application to quadratures, though it did contrive to indicate something of the scope of the general method of fluxions. With the transmission of the paper to John Collins in London, Newton's anonymity began to dissolve. Though a mediocre mathematician at best, Collins could recognize genius when he saw it. He received *De analysi* with the enthusiasm it deserved. Before he fulfilled Barrow's request and returned the paper, he took a copy. He showed the copy to others, and he wrote about the contents of the tract to a number of his correspondents: James Gregory in Scotland, René de

Sluse in the low countries, Jean Bertet and the Englishman Francis Vernon in France, G. A. Borelli in Italy, Richard Towneley and Thomas Strode in England. Years later, when Newton went through Collins papers, he was surprised to learn just how widely the paper had circulated. "Mr. Collins was very free in communicating to able Mathematicians what he had receiv'd from Mr. Newton . . .," he wrote anonymously in the supposedly impartial *Commercium epistolicum.* Meanwhile, Collins and Barrow wanted to publish it as an appendix to Barrow's forthcoming lectures on optics. This was more than Newton could contemplate. He drew back. A letter of Collins's indicates that they applied more suasion than a mere suggestion in passing; he thought Newton would "give way" eventually. He was mistaken. Newton withheld the publication of his method, the first episode in a long history of similar withdrawals. Thus quietly did Newton's apprehensions sow the seeds of vicious conflicts.

De analysi was not, however, devoid of effect on Newton's life. At the very time he communicated it to Barrow, Barrow was contemplating resignation from the Lucasian professorship of mathematics. The professorship, established scarcely five years earlier by the bequest of Henry Lucas, was the first new chair founded in Cambridge since Henry VIII had created the five Regius professorships in 1540. With the Adams professorship of Arabic, established in 1666, it brought the number of similar positions in the university to eight. It was the only one concerned in any way with mathematics and natural philosophy, which were otherwise hardly touched upon by the curriculum. By existing standards, Lucas endowed it magnificently; with its stipend of £100, more or less, from the income of lands purchased in Bedfordshire, it ranked behind the masterships of the great colleges and the two chairs in divinity (which were usually occupied by college masters) as the ripest plum of patronage in an institution much concerned with patronage. On 29 October 1669, this plum fell into the lap of an obscure young fellow of peculiar habits, apparently without connections, in Trinity College.

There are various stories about Barrow's resignation and Newton's appointment. One version holds that Barrow recognized his master in mathematics and resigned in his favor. Frankly, it is quite impossible to square this account with the features of life in the Restoration university as we know them. Another more recent one, more in harmony with the times, suggests that Barrow was angling for a higher position. It is known, I think, beyond doubt that Barrow was a man ambitious for preferment.

One has only to recall his invariable contributions to the valedictory volumes published by the university – not to mention their length! – to realize as much. Within a year of his resignation, he was appointed chaplain to the king, and within three years master of the college. Nevertheless, no rule demanded he resign the Lucasian chair before he courted further preferment, or for that matter that he forgo the ubiquitous royal dispensation to enjoy both at once, and it is hard to leave the third account of his resignation wholly out of the picture. In his own eyes, Barrow was a divine, not a mathematician; he resigned to devote himself to his true calling. Seventeenth-century society being what it was (that is, not wholly incommensurable with twentieth-century society), this motive was in no way incompatible with the other. Contemporary comments agreed in the assertion that Barrow effectively appointed Newton. Collins understood as much, and a generation later Conduitt did also.

According to the statutes, the Lucasian professor was required to read and expound "some part of Geometry, Astronomy, Geography, Optics, Statics, or some other Mathematical discipline" each week during the three academic terms, and each year he had to deposit in the university library copies of ten of the lectures he had read. In Restoration Cambridge, performance tended to diverge, often wildly, from statutory requirements. As far as the students were concerned, the fresh burden of lectures imposed upon them was only another item on a list now universally ignored. By 1660, college tutoring had virtually completed its conquest of university instruction. Barrow had complained of the neglect of his lectures when he was professor of Greek, and other professors echoed the story. Although we have no information about Newton's early experience, we do know what Humphrey Newton found upon his arrival fifteen years later. When Newton lectured, he recalled, "so few went to hear Him, & fewer y^t understood him, y^t oftimes he did in a manner, for want of Hearers, read to ye Walls." One of the fundamental facts of Newton's tenure as Lucasian professor of mathematics is the paucity of reference to his teaching. For forty years after 1687, he was the most famous intellectual in England, and there was every incentive for former students at the university to recall their connections with him. Even William Whiston, who became his disciple and successor, could barely remember having heard him. As far as we know, only two others ever claimed to have been instructed by him.

The record indicates that Barrow had already reduced the requirement of lecturing from three terms to one; Newton acquiesced in that schedule. He delivered a course of lectures in the Lent term of 1670, soon after his appointment. Thereafter he gave a series during the Michaelmas term (or at any rate he deposited manuscripts with such dates on them) each year through 1687. After 1687, he succumbed to the prevailing mode and held the position as a sinecure for fourteen years, during five of which he was not even resident in Cambridge. The record of his absences from Trinity during the earlier period supports the conclusion that he lectured only one term per year. Although he did not leave frequently, when he did do so he was as apt to go during term time as during vacations. Indeed he left for two weeks in London less than a month after his appointment. Nor did he concern himself excessively with the requirement that he deposit copies of ten lectures each year. In all, he eventually deposited four manuscripts that purported to contain annual courses of lectures through 1687. Questions have been raised about all four, so that it is quite impossible to know for sure on what he lectured. It appears highly probable that the lectures through 1683 corresponded roughly to those in the deposited manuscripts. Beginning in 1684, he may have lectured on the *Principia*, but the deposited manuscripts were merely drafts of that work which he sent in as the easiest way to fulfill the requirement.

As the topic for his first course of lectures, Newton chose, not the subject of *De analysi*, indeed not mathematics at all, but optics. He had required external stimulus to compose *De analysi*. During the following two years, he devoted a fair amount of time to mathematics, but again under external stimulus. What he turned to of his own accord was optics and the theory of colors. His accounts recorded the purchase of three prisms some time after February 1668, probably during one of the summer fairs. His earliest surviving letter, dated 23 February 1669, described his first reflecting telescope and referred obliquely to his theory of colors. Thus there is reason to think that Newton had resumed his investigation of colors before his appointment, and that he chose to lecture on the topic then foremost in his mind. Two or three years earlier, he had sketched out his theory of colors. Now the problem seized him in earnest and would not release him until he had conquered it. Reaching back to his incomplete investigation of 1666, he now worked out the full implications of his

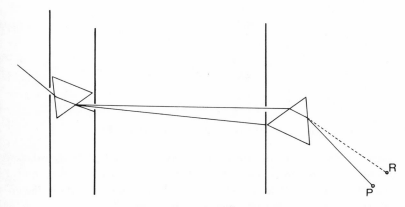

Figure 4. The *experimentum crucis*.

central insight and brought his theory of colors virtually to the form that he published more than thirty years later as his *Opticks*.

As he clarified the theory, so also he strengthened its experimental foundation. In 1666, he had begun, in a crude way, to employ a second prism to refract separate parts of the spreading spectrum. Now he refined the experiment into a form which could rigorously refute the theory of modification. He placed the second prism halfway across the room with its axis perpendicular to the first so that the full spectrum fell upon it. If, as the theory of modification might argue, dispersion as well as coloration was a modification introduced by the prism, the second prism ought to spread the spectrum into a square. Quite the contrary, it produced a spectrum inclined at an angle of 45 degrees. Further improvements suggested themselves. With the second prism set parallel to the first, Newton covered its face except for a small hole which admitted individual colors isolated from the rest of the spectrum, and he compared the quantities of their refractions. Finally he realized the importance of a demonstrably fixed angle of incidence on the second prism. To achieve this, he used two boards with small holes, one placed immediately beyond the first prism, the other immediately in front of the second. Because the boards were fixed in position, the two holes defined the path of the beam that fell on the second prism, also fixed in position beyond the second hole. (See Figure 4.) By turning the first prism slightly on its axis, Newton could

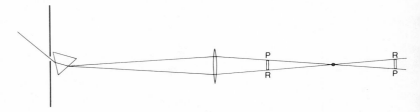

Figure 5. The reconstitution of white light with a lens.

transmit either end of the spectrum, pretty well if not perfectly isolated from the rest, into the second prism. There, as he expected, the blue rays were refracted more than the red. Neither beam suffered further dispersion. It was this experiment which Newton later called his *experimentum crucis.*

In 1669, Newton also greatly expanded his experimental demonstration that white is merely the sensation caused by a heterogeneous mixture of rays. To the brief experiments of 1666 in which he cast overlapping spectra on each other, he added one in which a lens collected a diverging spectrum and restored it to whiteness. If he intercepted the converging rays before the focus, he obtained an elongated spectrum reduced in size. At the focus, the spectrum disappeared into a white spot. On the far side of the focus, the spectrum reappeared with its order reversed. (See Figure 5.) Beyond the lens, no operation was performed on the light. When the spectrum merged into the focus, colors merged into whiteness. Because the individual rays retained their identity, colors reappeared as they separated anew beyond the focus. Newton was aware that impressions on the retina endure for about a second. The further thought occurred that all the elements of the heterogeneous mixture that produces the sensation of white need not be present at once. He mounted a wheel beyond the lens so that the heavy spokes intercepted individual colors of the converging spectrum. When he turned the wheel slowly, a succession of colors appeared at the focus. When he turned it fast enough that the eye could no longer distinguish the succession, white appeared once more.

As far as the theory of colors is concerned, the *Lectiones opticae,* which Newton probably composed late in 1669 and in 1670, concentrated on prismatic phenomena. They did not elaborate the explanation of the colors of solid bodies that Newton had sketched in his essay, "Of Colours." I assume, therefore, that the investigation that became the foundation for such an elaboration, contained in a paper with the title, "Of ye coloured circles twixt two contiguous glasses," dated from 1670 at the earliest.

Newton employed the technique suggested in 1666. He placed a lens of known curvature on a flat piece of glass, causing a series of colored rings to appear to him as he looked down on the apparatus. In 1666, he used a lens with a radius of curvature of 25 inches. Now he employed a lens with a radius of 50 feet, increasing the diameters of the rings nearly five times. What he demanded of his measurements tells us much about the man. Measuring with a compass and the unaided naked eye, he expected accuracy of less than one one-hundredth of an inch. With no apparent hesitation, he recorded one circle at 23½ hundredths in diameter and the next at 34⅓. When a small divergence appeared in his results, he refused to ignore it but stalked it relentlessly until he found that the two faces of his lens differed in curvature. The difference corresponded to a measurement of less than one one-hundredth of an inch in the diameter of the inner circle and about two one-hundredths in the diameter of the sixth. "Yet many times they imposed upon mee," he added grimly to his successful elimination of the error. Applying his measurements to the geometry of circles, Newton was able to establish the periodicity of the rings.

The paper on colored circles fit into many of Newton's interests. The difficulty of bringing a convex lens into contact with a flat sheet of glass impressed him; it appeared in all of his subsequent speculations as one of the key phenomena for understanding the nature of things. An experiment with water between the glasses instead of air gave him first-hand experience with capillary action, though earlier he had made notes in the "Quaestiones" about it from his reading. He made a film of water by letting a drop "creepe" between the glasses. From the varying results of two experiments, in one of which the glasses were pressed much harder, he concluded that the creeping in of the water altered the curvature of the glass because water has less incongruity with glass than air has. Congruity and incongruity were concepts he had met in Hooke's *Micrographia;* they too had a long history ahead of them in Newton's speculations. So did the aether to which he alluded several times, as he had done in the essay of 1666. Already at that time he had implied the principal features of his mechanical account of optical phenomena. It is not, he asserted, "yᵉ superficies of Glasse or any smoth pellucid body yᵗ reflects light but rather yᵉ cause is yᵉ diversity of Aether in Glasse & aire or in any contiguous bodys." He spoke of pulses in the aether in connection with thin films. References to pulses, which implied his mechanical explanation of the periodic rings also filled the paper on colored circles. The pulses were not light. Rather they were vibrations in the aether, set up by the blow of a corpuscle of light on the first surface of a film, which determined whether or not the corpuscle would be able to penetrate the second surface and thus be transmitted, or would be reflected.

Newton succeeded in establishing that the ratio of the pulses for purple, at one end of the spectrum, to those for red at the other was 9:14 or 13:20. This ratio remained the empirical foundation of Newton's quantitative treatment of colors in solid bodies. Bodies are composed of transparent particles the thickness of which determines the colors they reflect. He had demonstrated with the prism that ordinary sunlight is a heterogeneous mixture of rays, each with its own immutable degree of refrangibility. "And what is said of their refrangibility may be understood of their reflexibility; that is, of their dispositions to be reflected, some at a greater, and others at a less thickness of thin plates or bubbles, namely, that those dispositions are also connate with the rays, and immutable." Hence all the phenomena of colors derive from processes of analysis, whether refraction or reflection, which separate individual rays from the mixture. In 1666, Newton laid out the program and carried it through for refractions. Only about in 1670 did he fully work out the details for the colors of solid bodies.

With 1670, Newton's creative work in optics virtually came to an end. He had worked out the implications of his initial insight, answering to his own satisfaction the questions he had set himself. Though he would devote considerable time to the exposition of his theory, first in 1672, later in the 1690s, and carry out some minor experimentation, he had effectively exhausted his interest in the subject. Never again was it able to command his undivided attention.

᠀

During this same period, willy-nilly, Newton also did some work on mathematics. Two enthusiastic and persuasive men, Isaac Barrow and

John Collins, now knew his power and refused to let it rest. Barrow involved him in the publication of his two sets of lectures. Nor was the relation one-sided. Barrow allowed Newton to use his extensive mathematical library. Barrow also set him to mathematical tasks. In the fall of 1669, he suggested that Newton revise and annotate the *Algebra* of Gerard Kinckhuysen, which had recently been translated from Dutch into Latin. It was also Barrow who set him at work a year later on a revision and expansion of *De analysi*. The episodes from 1669 to 1671 constitute most of the known relationship between the two men. It was just as well for Newton's career that he chose to please the older scholar. Already Barrow had shown himself to be a powerful patron. Newton was to need his aid one more time.

John Collins proved a more pertinacious gadfly. He had been the ultimate source of the Kinckhuysen *Algebra*, which he had had translated from Dutch to supply the lack of a good introduction to the subject. It was out of the question that Collins should allow his new discovery to escape from his net of communication. Newton's exchange with Collins effectively introduces his surviving correspondence.

The "Observations on Kinckhuysen," which Newton completed and sent to Collins by the summer of 1670, served further to increase his fame within a limited circle of mathematicians. John Wallis, to whom Collins had not shown *De analysi* because of his reputation for plagiary, did hear about the annotations; he offered the opinion that Newton could bring them out as an independent treatise of his own. Towneley longed to see the Kinckhuysen volume "with those wonderfull additions of Mr Newton." James Gregory, a mathematician who approached Newton's stature, continued to correspond with Collins about Newton's method of expanding binomials into infinite series.

Whether he understood the full extent of the publicity or not, Newton no sooner sensed the consequences of Collins's adulation than his anxiety, lulled initially by the pleasure of recognition, began to mount anew. It was evident already in his letter of 18 February 1670. Collins had asked to publish a formula for annuities that Newton had sent. Newton agreed, "soe it bee wthout my name to it. For I see not what there is desirable in publick esteeme, were I able to acquire & maintaine it. It would perhaps increase my acquaintance, ye thing wch I cheifly study to decline." Already he had begun to fend off the suggestions that he publish *De analysi*. Now he began to withdraw from Collins's overeager embrace as well. He informed Collins in the letter of 18 February that he had found a way to compute the harmonic series with logarithms, but he did not include it because the calculations were "troublesom." Collins did not hear from him again until July.

When he finally sent Collins the "Observations on Kinckhuysen" in July, Newton accompanied them with a letter filled with defensive diffidence. He hoped he had done what Collins wanted. He left it entirely to Collins whether to print any or all of it. "For I assure you I writ w^t I send you not so much wth a designe y^t they should bee printed as y^t your desires should bee satisfied to have me revise ye booke. And so soone as you have read ye papers I have my end of writing them."

There remains [he added] but one thing more & thats about the Title page if you print these alterations wch I have made in the Author: For it may bee esteemed unhandsom & injurious to Kinck huysen to father a booke wholly upon him wch is soe much alter'd from what hee had made it. But I think all will bee safe if after ye words [nunc e Belgico Latine versa] bee added [et ab alio Authore locupletata.] or some other such note.

Not "enriched by Isaac Newton" but "enriched by another author"! Others might long to see his wonderful additions. Newton himself was primarily concerned that his name not appear.

Encouraged by receiving the annotations, Collins hastily wrote back that he noted Newton's agreement with him on the insufficiency of Kinckhuysen's treatment of surds. He sent along three books and asked Newton to pick out the best discussion of surds to insert in the volume. Rather wearily, Newton asked to have the manuscript back. It came at once with another letter filled with further questions and the promise of further publicity; "your paines herein," Collins assured him, "will be acceptable to some very eminent Grandees of the R Societie who must be made accquainted therewith . . . ." It was a clumsy thing to say to a man who had recently told him he studied chiefly to diminish his acquaintance. Over two months passed before Newton replied. On 27 September he informed Collins that he had thought some of composing a completely new introduction to algebra.

But considering that by reason of severall divertisements I should bee so long in doing it as to tire you patience wth expectation, & also that there being severall Introductions to Algebra already published I might thereby gain ye esteeme of one ambitious among ye croud to have my scribbles printed, I have chosen rather to let it passe w^{th}out much altering what I sent you before.

Collins never saw the manuscript again. He also heard no more from Newton for ten months.

Although he had misjudged Newton initially, Collins now realized that he was dealing with a man extraordinary in more ways than that of mathematical genius. He responded to Newton's silence with his own, and in December he described his relations with Newton to James Gregory. Gregory was eager to learn about Newton's general method of infinite series. Collins told him how Newton communicated individual series but not the general method, though he understood that he had written a treatise about it. Collins had sent him the annuity problem, hoping thereby to learn the general method. Newton had sent back only the formula; "hence observing a warinesse in him to impart, or at least an unwillingness to be at the paines of so doing, I desist, and doe not trouble him any more . . . ."

But in the end Collins could not deny his self-imposed mission and desist forever. In July 1671, he wrote a chatty letter about mathematics and the Kinckhuysen edition, which would have a better sale, he remarked, if it carried Newton's name. He also sent Newton a copy of Borelli's new book. Newton responded with deliberate incivility by suggesting that Collins not send him any more books. It would be sufficient if he merely informed him what was published. He did mention that he had intended to visit Collins on the occasion of the recent induction of the Duke of Buckingham as chancellor of the university, but a bout of sickness had prevented his making the trip to London. Somewhat grudgingly, it appears, he also added that he had reviewed his introduction to Kinckhuysen during the winter.

And partly upon Dr Barrows instigation, I began to new methodiz ye discourse of infinite series, designing to illustrate it wth such problems as may (some of them perhaps) be more acceptable then ye invention it selfe of working by such series. But being suddainly diverted by some buisinesse in the Country, I have not yet had leisure to return to those thoughts, & I feare I shall not before winter. But since you informe me there needs no hast, I hope I may get into ye humour of completing them before ye impression of the introduction, because if I must helpe to fill up its title page, I had rather annex something wch I may call my owne, & wch may bee acceptable to Artists as well as ye other to Tyros.

The new methodized discourse, known as the *Tractatus de methodis se-rierum et fluxionum (A Treatise of the Methods of Series and Fluxions)*, though Newton himself did not give it a title, was the most ambitious exposition of his fluxional calculus that Newton had yet undertaken. Drawing on both *De analysi* and the tract of October 1666, he produced an exposition of his method directed to that circle of mathematical artists with which he had communed so far only passively in the reading of their works.

For all its brilliance, the most remarkable thing about *De methodis* with its associated papers is the fact that Newton never completed it. From his letters, it appears that he started the treatise in the winter of 1670–1. A trip home in the spring interrupted him, and when he wrote to Collins on 20 July 1671, he said he had not returned to the papers and did not expect to before winter. "I hope I may get into ye humour of completing them . . . ." Because Newton was prone to similar comments, which were defensive maneuvers to ward off criticism by pretending lack of interest, we should pause before we take the comment seriously. The manuscript itself does seem to bear him out, however. It reveals an initial effort that came to a halt, a renewed attempt that advanced a bit further, and ulti-mate abandonment. In May 1672, Newton informed Collins that he had written the better half of the treatise the previous winter, but it had proved larger than he expected. It was not done; he might "possibly" complete it. By July he did not know "when I shall proceed to finish it." In fact, he never did.

No doubt the reluctance of London booksellers to publish mathe-matical books, which usually lost money, played a role in Newton's dilato-riness. It is impossible to assign the determining role to this factor, how-ever. The Royal Society subsidized the publication of Horrox's *Opera* in 1672. Edmund Gunter's *Workes* were republished the following year. In 1674, Barrow brought out a new edition of his lectures and in the years ahead proceeded with publications of Euclid's *Data* and *Elements* and of Archimedes and Apollonius. Other mathematical works, mostly elemen-tary ones to be sure, appeared continually. Had Collins ever gotten his hands on Newton's *De methodis*, he would have moved heaven and earth to put it into print as Edmond Halley would with another treatise fifteen years later. It was not the depression of the publishing trade; rather it was Newton who aborted the publication of a work which would have trans-formed mathematics. He never returned his annotations on Kinckhuysen to Collins, and he finally killed the edition by buying out the interest of the bookseller Pitts for four pounds. Collins never saw anything of the major treatise beyond the barest of tantalizing hints in a couple of letters. The unresolved, unresolvable tension that pulled Newton to and fro, as he responded to the warmth of praise, then fled in anxiety at the scent of criticism, worked now to suppress his masterful treatise.

For that matter, he was not terribly interested in it in any case. As he had told Collins, he was not in the humor to complete it. Nearly all of Newton's burst of mathematical activity in the period 1669–71 can be traced to external stimuli, to Barrow (armed initially with Mercator's work) and to Collins. His own interests had moved on. By 1675, Collins, who confessed that he had not heard from him for nearly a year, reported to Gregory that Newton was "intent upon Chimicall Studies and practises, and both he and Dr Barrow &c [were] beginning to thinke mathcall Speculations to grow at least nice and dry, if not somewhat barren . . . ."

૨ક

However, it proved to be impossible for Newton to retire again to the anonymity of his sanctuary. An irresistible current that would not let his gifts be hidden bore him forward. If not mathematics, then something else. Fittingly, it was a product of his hands rather than a child of his brain which brought the craftsman from Grantham fully into the view of the European scientific community. Though, contrary to the account long accepted, we now know that Newton's theory of colors did not lead him wholly to despair of refracting telescopes, he did build a reflecting telescope nevertheless. He cast and ground the mirror from an alloy of his own invention. He built the tube and the mount. And he was proud of his handiwork. He was still proud when he recalled it for Conduitt nearly sixty years later. "I asked him," Conduitt recorded, "where he had it made, he said he made it himself, & when I asked him where he got his tools said he made them himself & laughing added if I had staid for other people to make my tools & things for me, I had never made anything of it . . . ." The telescope was about six inches long, but it magnified nearly forty times in diameter, which, as Newton could be brought to admit, was more than a six-foot refractor could do. Later he made a second telescope. "When I made these," he confessed in the *Opticks*, "an Artist in London undertook to imitate it; but using another way to polishing them than I did, he fell much short of what I had attained to . . . ." The Lucasian professor was unable to restrain himself from proceeding on to lecture the artisans of London on the secrets of their craft.

Meanwhile he found it quite impossible not to show off his creation. His earliest surviving letter, of February 1669, is a description of it to an unknown correspondent written as a result of a promise to Mr. Ent, to whom he had presumably shown or mentioned the telescope. When he met Collins in London at the end of 1669, he told him about the telescope, allowing it in his account to magnify 150 times. He must have been showing it off in Cambridge. In December 1671, Collins repeated to Francis Vernon what Mr. Gale (a fellow of Trinity) had written of it from Cambridge. By January, Towneley was asking about it excitedly, and Flamsteed had heard of it both from London and from a relative who had recently been in Cambridge. Perhaps Collins had never informed the eminent grandees of the Royal Society about Newton's mathematical achievements, but they heard about the telescope all right and asked to see it late in 1671. At the very end of the year, Barrow delivered it to them.

When it arrived, the telescope caused a sensation. Early in January, Newton received a letter from Henry Oldenburg, the secretary of the society.

S^r

Your Ingenuity is the occasion of this addresse by a hand unknowne to you. You have been so generous, as to impart to the Philosophers here, your Invention of contracting Telescopes. It having been considered, and examined here by some of ye most eminent in Opticall Science and practise, and applauded by them, they think it necessary to use some meanes to secure this Invention from ye Usurpation of forreiners; And therefore have taken care to represent by a scheme that first Specimen, sent hither by you, and to describe all ye parts of ye Instrument, together wth its effect, compared wth an ordinary, but much larger, Glasse; and to send this figure, and description by ye Secretary of ye R. Soc. (where you were lately by ye Ld Bp. of Sarum [Seth Ward] proposed Candidat) in a solemne letter to Paris to M. Hugens, thereby to prevent the arrogation of such strangers, as may perhaps have seen it here, or even wth you at Cambridge; it being too frequent, y^t new Invention and contrivances are snatched away from their true Authors by pretending bystanders; But yet it was not thought fit to send this away wthout first giving you notice of it, and sending to you ye very figure and description, as it was here drawne up; y^t so you might adde, & alter, as you shall see cause; wch being done here w^th, I shall desire your favour of returning it to me wth all convenient speed, together wth such alterations, as you shall think fit to make therein . . . .

S^r

your humble servant

Oldenburg

True to its word, the Royal Society sent a description of the instrument to Huygens; it even sent a general account written on 1 January ahead, so concerned were its members to secure the credit to Newton. Huygens was no less pleased than they; he called it the "marvellous telescope of Mr. Newton . . . ." The Society employed Christopher Cock, an instrument-maker in London, to construct a reflecting telescope four feet in length, and later one of six feet, though both failed for want of satisfactory mirrors. Having nominated Newton, the society proceeded to his full election on 11 January.

The ritual dance performed with Collins now commenced anew. Newton fairly beamed as the warm glow of praise fell about him.

At the reading of your letter [he replied to Oldenburg] I was surprised to see so much care taken about securing an invention to mee, of wch I have hitherto had so little value [sic!]. And therefore since the R. Society is pleased to think it worth the patronizing, I must acknowledg it deserves much more of them for that, then of mee, who, had not the communication of it been desired, might have let it still remained in private as it hath already done some yeares.

Despite his pretense of indifference, Newton picked up the society's intention to send a description to Huygens and suggested that it be sure he realized that the telescope eliminated colors from the image. He volunteered instructions about its maintenance, and in his next two letters he sent information about alloys he had tried for mirrors. He readily agreed to the publication of the description without even suggesting that his name be withheld.

I am very sensible of the honour done me by ye B^p of Sarum in proposing mee Candidate [he concluded his initial reply to Oldenburg] & wch I hope will bee further conferred upon mee by my Election into the Society. And if so, I shall endeavour to testify my gratitude by communicating what my poore & solitary endeavours can effect towards ye promoting your Philosophicall designes.

The Royal Society could not have guessed that the final sentence contained a hidden promise. Newton unveiled it on 18 January. He informed the society that "I am purposing them, to be considered of &

examined, an accompt of a Philosophicall discovery wch induced mee to the making of the said Telescope, & wch I doubt not but will prove much more gratefull then the communication of that instrument, being in my Judgment the oddest if not the most considerable detection wch hath hitherto beene made in the operations of Nature." The ritual dance had further figures, however. As he had found already with Collins, it proved to be not quite that simple to divulge his discovery. A week and a half later, he had not yet sent it, and he felt compelled to go through the remaining steps. He wrote that he hoped he could "get some spare howers" to send off the account. Wickins needed the spare hours as much as Newton, who set him to work copying the paper. The die had been cast, however. It was too late to withdraw. On 6 February 1672, Newton finally mailed an account of his theory of colors to London. For the moment the positive pole prevailed. Swept along by the success of his telescope, Newton stepped publicly into the community of natural philosophers to which he had hitherto belonged in secret.

5

Publication and Crisis

THE PAPER ON COLORS that Newton sent to the Royal Society early in 1672 in the form of a letter addressed to Henry Oldenburg did not contain anything new from Newton's point of view. The occasion provided by the telescope had come at an opportune time. At Barrow's behest, Newton had been revising Barrow's lectures for publication during the winter. He had not found it a great chore to produce a succinct statement of his own theory buttressed by three prismatic experiments that he took to be most compelling. He thought it relevant to include a special discussion of how the discovery had led him to devise the reflecting telescope. The continuing correspondence provoked by the initial paper, which intruded intermittently on his time and consciousness during the following six years, also involved only one addition to his optics, his introduction to diffraction and brief investigation of it. Aside from diffraction, the entire thrust of his concern with optics during the period was the exposition of a theory already elaborated.

The continuing discussion forced Newton to clarify some issues. When he wrote in 1672, he had not yet fully separated the issue of heterogeneity from his corpuscular conception of light, and he allowed himself to assert that, because of his discovery, it could "be no longer disputed . . . whether Light be a Body." He could hardly have been more mistaken. Within a week of the paper's presentation, Robert Hooke produced a critique that mistook corpuscularity for its central argument and proceeded to dispute it with some asperity. The lesson was not wasted. Though he continued to believe in the corpuscular conception, Newton learned to insist that the essence of his theory of colors lay in heterogeneity alone. This was a matter of clarification and exposition, however, not an alteration of his theory. The very fact that six years of discussion effected no change, such that his *Opticks,* finally published in 1704, merely restated conclusions worked out in the late 1660s, testifies to the intensity and rigor of the early investigation.

The discussion that followed on the paper of 1672 tells us less about optics than about Newton. For eight years he had locked himself in a remorseless struggle with Truth. Genius of Newton's order exacts a toll. Eight years of uneaten meals and sleepless nights, eight years of continued ecstasy as he faced Truth directly on grounds hitherto unknown to the human spirit, took its further toll. And exasperation that dullness and stupidity should distract him from the further battles in which he was already engaged on new fields added the final straw. By 1672, Newton had lived with his theory for six years, and it now seemed obvious to him. For everyone else, however, it still embodied a denial of common sense that made it difficult to accept. Their inability to recognize the force of his demonstrations quickly drove Newton to distraction. He was unprepared for anything except immediate acceptance of his theory. The continuing need to defend and explain what he took to be settled plunged him into a personal crisis.

To be sure, the initial response gave no hint of the crisis to follow. Almost before the ink had dried on his paper of 6 February, Newton received a letter from Oldenburg. Filled with lavish praise, it informed him that his paper had been read to the Royal Society, where it "mett both with a singular attention and an uncommon applause . . . ." The society had ordered it to be printed forthwith in the *Philosophical Transactions* if Newton would agree. The tension caused by the decision to send the paper can be heard in Newton's relief as he read Oldenburg's letter.

I before thought it a great favour to have beene made a member of that honourable body; but I am now more sensible of the advantage. For beleive me S^r I doe not onely esteem it a duty to concurre wth them in ye promotion of reall knowledg, but a great privelege that instead of exposing discourses to a prejudic't & censorious multitude (by wch means many truths have been bafled & lost) I may wth freedom apply my self to so judicious & impartiall an Assembly.

He assented to the publication of the paper with only a slight – for him, compulsory – demur.

Accordingly, the paper appeared in the *Philosophical Transactions* for 19 February 1672. Together with the description of his telescope, which the following issue carried, it established Newton's reputation in the world of natural philosophy. Oldenburg took care to publicize both items in his extensive correspondence with natural philosophers throughout Europe. The replies he received indicate that both were noticed. The telescope caught the eye of leading astronomers everywhere – Cassini, Auzout, and Denis in Paris, and Hevelius in Danzig. Oldenburg specifically called the paper on colors to Huygens's attention when he mailed the *Philosophical Transactions* to him. Huygens replied that "the new Theory . . . appears very ingenious to me." To be sure, Huygens later expressed reservations about the theory; meanwhile, in April, Newton received what can only have appeared as praise from the recognized leader of European science. A young English astronomer, John Flamsteed, who would soon become the first Astronomer Royal, commented on the paper, though without much comprehension. A young German savant resident in Paris, Gottfried Wilhelm Leibniz, then unknown but as determined to make his way in natural philosophy as he was destined to, indicated that he had seen it. Towneley reported to Oldenburg that Sluse had asked him to translate it into French so that he might read it. For himself, Towneley found the paper "so admirable" that he urged the publication of a Latin translation for the benefit of philosophers across Europe. As a result of the telescope and the paper on colors, Newton soon found himself the recipient of presentation copies of books by Huygens and Boyle. Never again could he return to the anonymity of the early years in Cambridge. Once and for all, he had installed himself in the community of European natural philosophers, and among its leaders.

Newton did not see every comment on his theory of colors that Oldenburg and others received. He did see enough that he should have been gratified by its overwhelmingly favorable reception. The praise was not unanimous, however. Newton had concluded the paper with a seeming invitation to comment and criticism: "That, if any thing seem to be defective, or to thwart this relation, I may have an opportunity of giving further direction about it, or of acknowledging my errors, if I have committed any." Alas, within two weeks he received a lengthy critique from Robert Hooke, the established master of the subject in England, a condescending commentary that contrived to imply that Hooke had performed all of Newton's experiments himself while it denied the conclusions Newton drew from them. Initially, Newton chose to ignore Hooke's tone.

I received your Feb 19th [Newton wrote to Oldenburg]. And having considered Mr Hooks observations on my discourse, am glad that so acute an objecter hath said nothing that can enervate any part of it. For I am still of the same judgment & doubt not but that upon severer examinations it will bee found as certain a truth as I have asserted it. You shall very suddenly have my answer.

The critique must have rankled more than he let on, however. Instead of receiving the answer suddenly, Oldenburg had to wait three months; and when it arrived, its tone was rather less unruffled.

Meanwhile other comments and critiques arrived. Sir Robert Moray, the first president of the Royal Society, proposed four experiments (which betrayed no understanding of the question) to test the theory. More significant were the objections of the French Jesuit Ignace Gaston Pardies, professor at the Collège de Louis-le-Grand and a respected member of the Parisian scientific community. He pointed out that for certain positions of the prism the sine law of refraction could account for the diverging spectrum because all the sun's rays were not incident on the prism's face at the same angle; and he questioned the *experimentum crucis* on the same grounds of unequal incidence. In fact, Newton's initial paper had adequately answered both objections. Nevertheless, Pardies's letter was an intelligent comment by a man obviously knowledgeable in optics. It was also respectful in tone, though Pardies made the mistake of opening with a reference to Newton's "very ingenious Hypothesis . . . ." Hooke had also called the theory of colors Newton's "hypothesis" several times. Now he began to bridle.

I am content [he concluded his reply to Pardies, manifestly discontent] that the Reverend Father calls my theory an hypothesis if it has not yet been proved to his satisfaction. But my design was quite different, and it seems to contain nothing else than certain properties of light which, now discovered, I think are not difficult to prove, and which if I did not know to be true, I should prefer to reject as vain and empty speculation, than acknowledge them as my hypothesis.

Pardies did not propose to start a quarrel. He apologized handsomely and accepted Newton's explanation of why the unequal incidence of the sun's rays on the prism could not explain the divergence of the spectrum. He raised a further question, however: Could not Grimaldi's recent discovery, diffraction, explain the divergency?

In answer to this [Newton replied], it is to be observed that the doctrine which I explained concerning refraction and colours, consists only in certain properties of light, without regarding any hypotheses by which those properties might be explained. For the best and safest method of philosophizing seems to be, first to enquire diligently into the properties of things, and to establish those properties by experiments and then to proceed more slowly to hypotheses for the explanation of them. For hypotheses should be employed only in explaining the properties of things, but not assumed in determining them; unless so far as they may furnish experiments. For if the possibility of hypotheses is to be the test of the truth and reality of things, I see not how certainty can be obtained in any science; since numerous hypotheses may be devised, which shall seem to overcome new difficulties. Hence it has been here thought necessary to lay aside all hypotheses, as foreign to the purpose . . . .

As the rest of the correspondence would further demonstrate, the discussion of colors provided Newton with his first serious occasion to explore questions of scientific method. Pardies expressed himself satisfied with the additional explanations that Newton offered, though there is no evidence that he accepted the theory.

During all this time, Hooke's critique of the February paper and the need to reply hung over Newton's head. Hooke and Newton were probably fated to clash. Newton had conceived his theory of colors in reaction to Hooke's. For his part, Hooke considered himself the authority on optics and resented the appearance of an interloper. When Newton's telescope set the Royal Society agog, he submitted a memorandum about a discovery using refractions that would perfect optical instruments of all sorts to the limit anyone could desire, far beyond Newton's invention. Unfortunately, he concealed the discovery itself in a cipher. He approached the paper on colors in much the same way, with a magisterial tone of authority, which would have been galling to a person less sensitive than Newton. Two scientists more different are hard to imagine. Though highly gifted, Hooke was more plausible than brilliant. He had ideas on every subject and was ready to put them into print without much hesitation. Newton in contrast was obsessed with the ideal of rigor and could hardly convince himself that anything was ready for publication. Hooke later confessed that he spent all of three or four hours composing his observations on Newton's paper. He had cause to regret his haste. Newton spent three months on his response. It may be relevant as well that Hooke was sick enough with consumption that later in the year he was not expected to survive.

Hooke submitted his critique to the Royal Society on 15 February, one week after Newton's paper was read. Newton had a copy by 20 February. Hooke granted Newton's experiments, "as having by many hundreds of tryalls found them soe," but not the hypothesis by which he explained them. "For all the expts & obss: I have hitherto made, nay and even those very expts which he alledged, doe seem to me to prove that light is nothing but a pulse or motion propagated through an homogeneous, uniform and transparent medium: And that Colour is nothing but the Disturbance of y^t light . . . by the refraction thereof . . . ." The burden of Hooke's critique was the reassertion of his own version of the modification theory as he had published it in *Micrographia*. He protested as well against Newton's abandonment of refracting telescopes. "The truth is, the Difficulty of Removing that inconvenience of the splitting of the Ray and consequently of the effect of colours, is very great, but not yet insuperable." He had already overcome it in microscopes, he asserted, but had been too busy to apply his discovery to telescopes. Like a true mechanical philosopher, Hooke kept returning to picturable images such as split rays to express his theory of colors. He saw Newton's theory in similar terms, as primarily an exposition of the corpuscular hypothesis, and he assured Newton that he would solve the phenomena of light and colors not only by his own hypothesis but by two or three others as well, all different from Newton's. He failed entirely to come to grips with Newton's experimental demonstration of the fact of heterogeneity.

Although Newton initially promised to reply at once, he planned an answer that required more time. Whatever else Hooke's critique contained, it reasserted the modification theory of colors without compromise. Newton decided to seize the opportunity it offered for a fully elaborated exposition of his own theory of analysis. He drew heavily on his *Lectiones opticae* for experimental support that he had omitted in his brief initial paper. Nor did he stop there. He composed as well an exposition of the phenomena of thin films as they pertained to the colors of bodies and the heterogeneity of light, more than a first draft of the "Discourse of Observations" of 1675 and Book II of the *Opticks*, because extensive passages appeared verbatim as they would be published thirty years hence. What Newton drafted in the early months of 1672 was a treatise on optics which contained a sometimes briefer exposition of all the elements of his ultimate work except Book II, Part IV (the phenomena of thick plates), Book III (his brief exposition of diffraction), and the Queries. Because it included the first sketch of his "Hypothesis of Light" (1675), it did contain material analogous to some of the Queries. Published in 1672, the small treatise would have advanced the science of optics by thirty years.

It was not published, however. In March, he told Oldenburg he had not yet completed it, and in April he delayed again. Perhaps by now he was looking for an excuse not to send it. Two years earlier, he had refused to have his name attached to a formula for annuities lest it increase his acquaintance, which he studied chiefly to diminish. The telescope and the paper on colors had shown how right he had been. By early May, four months after he sent the telescope to London, he had received twelve letters and written eleven answers about the telescope and colors – hardly a crushing burden, but not a decrease in acquaintance either. Discussing the events of the spring four years later, Newton told Oldenburg that "frequent interruptions that immediately arose from the letters of various persons (full of objections and of other matters) quite deterred me from the design [of publishing the *Optical Lectures*] and caused me to accuse myself of imprudence, because, in hunting for a shadow hitherto, I had sacrificed my peace, a matter of real substance." If Newton was looking for an excuse, Oldenburg's letter of 2 May offered one. Oldenburg urged him to omit Hooke's and Pardies's names from his answers and to deal with their objections alone, "since those of the R. Society ought to aime at nothing, but the discovery of truth, and ye improvem^t of knowledge, and not at the prostituting of persons for their mis-apprehensions or mis-takes." Another statement in the letter, that "some begin to lay more weight upon [your theory of light] now, than at first," may have served to increase his unhappiness with the suggestion. Had Oldenburg misled him about the reception of his paper? Initially, Newton acquiesced in the request, though it clearly irritated him. He was more than irritated after he had brooded over it for two weeks.

I understood not your desire of leaving out Mr Hooks name, because the contents would discover their Author unlesse the greatest part of them should be omitted & the rest put into a new Method wthout having any respect to ye Hypothesis of colours described in his *Micrographia*. And then they would in effect become new objections & require another Answer then what I have written. And I know not whether I should dissatisfy them that expect my answer to these that are already sent to me.

He had decided, he continued, not to send all that he had prepared, though he still intended to include a discourse on "the Phaenomena of Plated Bodies," in which he showed that rays differ in reflexibility as well as in refrangibility and related the colors of bodies to the thickness of their particles.

A few days later, in a state of great agitation, he wrote to Collins, thanking him for his offer to undertake the publication of his optical lectures.

But I have now determined otherwise of them; finding already by that little use I have made of the Presse, that I shall not enjoy my former serene liberty till I have done with it; wch I hope will be so soon as I have made good what is already extant on my account.

He could not put the subject out of his mind, and after a paragraph about his mathematical work he returned to it.

I take much satisfaction in being a Member of that honourable body the R. Society; & could be glad of doing any thing wch might deserve it: Which makes me a little troubled to find my selfe cut short of that fredome of communication wch I hoped to enjoy, but cannot any longer without giving offense to some persons whome I have ever respected. But tis no matter, since it was not for my own sake or advantage y^t I should have used that fredome.

When he finally sent the reply on 11 June, Newton had eliminated the discourse on thin films along with most of the material from his *Lectiones*. What he did send was an argument pointed to the issue of analysis versus modification. Though not so well known as the paper of February, the answer to Hooke supplemented it brilliantly in the use of prismatic phenomena to support the theory of colors. Equally, it presented an argument *ad hominem*. Far from omitting Hooke's name, Newton inserted it in the first sentence of the reply, in the last, and in more than twenty-five others in between. He virtually composed a refrain on the name Hooke. Successive drafts of various passages progressed through three and four stages, each one more offensive than the last.

I must confesse [he said in the final version of the opening paragraph] at ye first receipt of those Considerations I was a little troubled to find a person so much concerned for an *Hypothesis*, from whome in particular I most expected an unconcerned & indifferent examination of what I propounded. . . . The first thing that offers itselfe is lesse agreable to me, & I begin with it because it is so. Mr Hook thinks himselfe concerned to reprehend me for laying aside the thoughts of improving Optiques by *Refractions.* But he knows well y^t it is not for one man to prescribe Rules to ye studies of another, especially not without understanding the grounds on wch he proceeds.

After setting Hooke straight on that score, Newton turned to Hooke's considerations of his theory.

And those consist in ascribing an Hypothesis to me wch is not mine; in asserting an Hypothesis wch as to ye principall parts of it is not against me; in granting the greatest part of my discourse if explicated by that Hypothesis; & in denying some things the truth of wch would have appeared by an experimentall examination.

Newton showed Hooke how to reconcile the wave hypothesis to his theory of colors, before he offered the opinion that Hooke's theory was "not onely *insufficient,* but in some respects *unintelligible.*" He even instructed Hooke, who had boasted that he could perfect optical instruments in general, how to improve microscopical observations, Hooke's special province, by the use of monochromatic light. So much for Hooke's pretended perfection of refracting instruments!

In a covering letter of the same date to Oldenburg, Newton assumed that Hooke would find nothing objectionable in the reply because he had avoided "oblique & glancing expressions . . . ." That is, he employed the broadsword instead of the rapier. Where Hooke's observations had been irritatingly patronizing, Newton's reply was viciously insulting – a paper filled with hatred and rage. It established for his relations with Hooke a pattern that was never broken. The Royal Society forebore to print Hooke's critique lest it appear disrespectful to Newton. It did allow Hooke to endure the humiliation, first of hearing the response read at a meeting, then of seeing it in print in the *Philosophical Transactions.*

Hooke was not the sole cause of Newton's exasperation. The entire exchange, the need to amplify and explicate what seemed perfectly obvious, annoyed him. On 19 June, he asked Oldenburg that he "not yet print any thing more concerning the Theory of light before it hath been more fully weighed." On 6 July, he requested again that Pardies's second letter not be published, though he relented when Oldenburg told him it was already at the printer. In the letter of 6 July, he tried to restate the question in a form that would terminate discussion.

I cannot think it effectuall for determining truth to examin the severall ways by wch Phaenomena may be explained, unless there can be a perfect enumeration of all those ways. You know the proper Method for inquiring after the properties of things is to deduce them from Experiments. And I told you that the Theory wch I propounded was evinced to me, *not by inferring tis thus because not otherwise,* that is not by deducing it onely from a confutation of contrary suppositions, but *by deriving it from Experiments concluding positively & directly.* The way therefore to examin it is by considering whether the experiments wch I propound do prove those parts of the Theory to wch they are applyed, or by prosecuting other experiments wch the Theory may suggest for its examination.

He proceeded then to reduce his theory to eight queries that could be answered by experiments. Let all objections from hypotheses be withheld. Either show the insufficiency of his experiments or produce other experiments that contradict him. "For if the Experiments, wch I urge be defective it cannot be difficult to show the defects, but if valid, then by proving the Theory they must render all other Objections invalid."

Newton obviously meant that the experiments he had sent already answered his eight queries. Unfortunately the Royal Society directed that experiments be performed to test them, and Oldenburg, with all the delicacy of an uncoordinated cow, asked Newton to suggest some. It was 21 September before he brought himself to reply, and then only to say he was busy with other things. Oldenburg heard no more that autumn. Neither did Newton's other correspondent, Collins, until he received a lengthy commentary on Gregory's remarks about telescopes in December. Newton explained to Collins that he had written such a "long scribble . . . because Mr Gregory's discours looks as if intended for the Press." Finally, in January, Oldenburg succeeded in eliciting from Newton a reply to a query on the improbable subject of cider ("wch liquor I wish, wth you, propagated far and near in England . . ."). Indeed, cider later became one of their staples of correspondence, a subject free of emotional investment, whatever its content of spirit.

Having not understood the silence, Oldenburg immediately followed Newton's response by forwarding a new critique, from no less a figure than Huygens. This was the fourth comment Newton had received from Huygens, each one less enthusiastic than the one before. When the paper appeared, Huygens found it "very ingenious." In the summer, it still seemed "very probable" to him, though he doubted what Newton said about the magnitude of chromatic aberration. Newton sent a brief explication. By autumn Huygens thought things could be otherwise than the theory held and suggested that Newton be content to let it pass as a very probable hypothesis. "Moreover, if it were true that from their origin some rays of light are red, others blue, etc., there would remain the great difficulty of explaining by the mechanical philosophy in what this diversity of colors consists." Oldenburg forwarded the comment to Newton; Newton did not answer. Of all the natural philosophers in Europe, Huygens was subjecting Newton's theory to its most searching scrutiny. In January 1673, he sent his fourth and fullest comment. It was also his most critical.

I have seen, how Mr. Newton endeavours to maintain his new Theory concerning Colours. Me thinks, that the most important Objection, which is made against him by way of Quaere, is that, Whether there be more than two sorts of Colours. For my part, I believe, that an Hypothesis, that should explain mechanically and by the nature of motion the Colors Yellow and Blew, would be sufficient for all the rest, in regard that those others, being only more deeply charged (as appears by the Prismes of Mr. Hook) do produce the dark or deep-Red and Blew; and that of these four all the other colors may be compounded. Neither do I see, why Mr. Newton doth not content himself with the two Colors, Yellow and Blew; for it will be much more easy to find an Hypothesis by Motion, that may explicate these two differences, than for so many diversities as there are of other Colors. And till he hath found this Hypothesis, he hath not taught us, what it is wherein consists the nature and difference of Colours, but only this accident (which certainly is very considerable,) of their different Refrangibility.

Once again, the mechanical philosophy with its demand for picturable explanatory images obstructed the understanding of Newton's discovery that light is heterogeneous.

Newton waited two more months to respond, only to indicate then that Huygens's private letter to Oldenburg did not call for an answer from him. If Huygens expected an answer, however, and intended "y^t they should be made publick," he would do so if he had Huygens's agreement "y^t I may have liberty to publish what passeth between us, if occasion be." In case that were not curt enough, he added something else for Oldenburg.

S^r I desire that you will procure that I may be put out from being any longer fellow of ye R. Society. For though I honour that body, yet since I see I shall neither profit them, nor (by reason of this distance) can partake of the advantage of their Assemblies, I desire to withdraw.

In regard to the threat of withdrawal, Oldenburg expostulated with Newton briefly and offered to have him excused from "ye trouble of sending hither his qterly payments & without any reflection." Newton, who was seeking to avoid complications, not to multiply them, did not pursue the issue, and it simply passed. In April, he sent to Huygens a reply which returned again to the issue of explanatory hypotheses. He could not rest satisfied with two colors because experiments showed that other colors are equally primary and cannot be derived from yellow and blue. Nor was it easier to frame a hypothesis for only two "unless it be easier to suppose that there are but two figures sizes & degrees of velocity or force of the aetherial corpuscles or pulses rather then an indefinite variety, wch certainly would be a very harsh supposition." No one is surprised that the waves of the sea and the sand on the shore reveal infinite variety. Why should the corpuscles of shining bodies produce only two sorts of rays?

But to examin how colours may be thus explained Hypothetically is besides my purpose. I never intended to show wherein consists the nature and difference of colours, but onely to show that *de facto* they are originall & immutable qualities of the rays wch exhibit them, & to leave it to others to explicate by Mechanicall Hypotheses the nature & difference of those qualities; wch I take to be no very difficult matter.

He went on to discuss the other issues Huygens had raised; and though he avoided the deliberately insulting tone of his reply to Hooke, he could not conceal his vehemence. Certainly Oldenburg did not miss it. "I can assure you," he wrote to Huygens, "that Mr. Newton is a man of great candor, as also one who does not lightly put forward the things he has to say." Huygens, who was not used to being addressed as a delinquent schoolboy, did not miss it either; "seeing that he maintains his doctrine with some warmth," he replied, "I do not care to dispute." He did permit himself a few pointed comments in a tone of icy hauteur. With one more letter from a somewhat chastened Newton, the exchange came to an end. Though Huygens had ample excuse to take offense, he recognized the quality of his opponent and chose not to. The very letter that carried his response enclosed a list of English scientists to whom Oldenburg should present copies of his newly published *Horologium oscillatorium;* Newton was among them. What is more to the point, Huygens allowed himself to be convinced, even though the heterogeneity of light posed difficulties, which he never surmounted, to the specific form in which he couched his wave theory of light.

Oldenburg had mentioned Newton's threat to withdraw from the Royal Society to Collins, who in turn commented on it to Newton.

I suppose there hath been done me no unkindness [Newton wrote him in May], for I met wth nothing in y^t kind besides my expectations. But I could wish I had met with no rudeness in some other things. And therefore I hope you will not think it strange if to prevent accidents of that nature for ye future I decline that conversation wch hath occasioned what is past.

When Collins showed him this, Oldenburg asked Newton to "passe by the incongruities" committed against him by members of the Royal Society. After all, every assembly had members who lacked discretion.

The incongruities you speak of, I pass by [Newton told him]. But I must, as formerly, signify to you, y^t I intend to be no further sollicitous about matters of Philosophy. And therefore I hope you will not take it ill if you find me ever refusing doing any thing more in y^t kind, or rather y^t you will favour me in my determination by preventing so far as you can conveniently any objections or other philosophicall letters that may concern me.

Oldenburg did not receive another letter from Newton for eighteen months.

Collins also found his correspondence interrupted. In the summer of 1674, Newton acknowledged the receipt of a book on gunnery and even commented on its content. "If you should have occasion to speak of this to ye Author," he added, "I desire you would not mention me becaus I have no mind to concern my self further about it." Late in 1675, Collins told Gregory that he had neither seen nor written to Newton for a year, "not troubling him as being intent upon Chimicall Studies and practices, and both he and Dr Barrow &c beginning to thinke mathcall Speculations to grow at least nice and dry, if not somewhat barren." Collins's correspondence with Newton never revived.

Oldenburg and Collins had functioned as Newton's contact with the learned world outside Cambridge. Although ample opportunities to correspond directly with men of the caliber of Gregory and Huygens had presented themselves, Newton had refused to grasp them. He communi-

cated with others through the two intermediaries, who virtually monopolized his correspondence. In cutting their access to him, Newton attempted to regain his former solitude. After the publications of 1672,
however, that was impossible. Huygens's presentation copy of the *Horologium* demonstrated as much, and in September 1673 Boyle confirmed
the point by presenting him a copy of his book on effluvia. Nevertheless,
for the moment, a modicum of criticism had sufficed, first to incite him to
rage, and then to drive him into isolation.

ঽঙ

In 1676, a mathematical correspondence that Oldenburg carried on with
the aid of Collins as part of his program of philosophical communication
spilled over to include Newton. The correspondence dated back beyond
the early months of 1673, when a young German philosopher, Gottfried
Wilhelm Leibniz, visited the Royal Society. At the beginning of 1673,
Leibniz was still very much a tyro in mathematics, but he was advancing
with giant strides toward its leading ranks. He made mathematics the
focus of a correspondence with Oldenburg, initiated during a visit to
London, when he was elected to membership in the Royal Society.
Oldenburg, who was not a mathematician, pressed Collins into service to
support his end of the exchange. Collins, of course, had made it his
business to stay in touch with the leaders of British mathematics, especially Gregory and Newton.

Leibniz achieved the fundamental insights of his differential calculus,
which was virtually identical to Newton's fluxional method, during the
autumn of 1675. He developed his distinctive notation, in which the
calculus still expresses itself, at that time. All of this has been established,
not from Leibniz's assertions but from his manuscripts, just as Newton's
invention of the fluxional method has been. At the end of 1675, it is
doubtful that Newton was aware of Leibniz's existence, though he may
have heard his name spoken at the Royal Society earlier that year in
connection with a bitter exchange between Oldenburg and Hooke about
Huygens's spring-driven watch. To the best of our knowledge, Newton
was also unaware that reports on his mathematical achievements, with
materials from his letters and from *De analysi,* were being sent to Leibniz
by Collins through Oldenburg. For this, he alone was to blame. He had
consistently discouraged communication and cut himself off when others

were eager to discuss and learn. Years later, after a bitter priority dispute had broken out, when Newton learned what Collins had sent, he drew his own sinister conclusions. What is clear from the correspondence, however, is that by the end of 1675, the critical period in Leibniz's own development, he had received only some of Newton's results without demonstrations, and that these results had been confined to infinite series. To be sure, infinite series formed an integral part of the fluxional method, but Leibniz had not heard of their broader ramifications.

In 1676, Newton both learned who Leibniz was and entered the correspondence himself. Leibniz wrote to Oldenburg in May asking for demonstrations of two series. Both Oldenburg and Collins urged Newton to respond. The request came at an unwelcome time. A new correspondence that challenged his theory of colors had opened, and Newton was allowing it to agitate him excessively. Nevertheless, he acceded to the request, and on 13 June 1676 he completed a letter for Leibniz. Once again, he chose not to enter into direct communication. He addressed the letter to Oldenburg, who forwarded a copy to Leibniz on 26 July.

Newton wrote two letters for Leibniz in 1676. Nearly forty years later, he cited them as evidence against Leibniz in the priority dispute and labeled them the earlier letter and the later letter, the *Epistola prior* and the *Epistola posterior*. Replying in the first to Leibniz's question about the foundation of the two series, he drew upon the combined reservoir of *De analysi* and *De methodis* to present a general exposition of series, which included his binomial theorem and illustrations of its use. If Leibniz had hitherto considered Newton merely as one among a number of English mathematicians, the *Epistola prior* disabused him. Nor was he reluctant to express his admiration. "Your letter," he wrote to Oldenburg immediately upon its receipt, "contains more numerous and more remarkable ideas about analysis than many thick volumes published on these matters . . . . Newton's discoveries are worthy of his genius, which is so abundantly made manifest by his optical experiments and by his catadioptrical tube [the reflecting telescope]." He went on to show Newton that he knew a thing or two about infinite series himself, to expound his general method of transformations, as he called it, and to put some specific questions to Newton.

Leibniz's new questions provided the occasion of the *Epistola posterior*. Before Newton could write it, however, Leibniz himself visited London for ten days in October. While he was there, he conversed with Collins, who opened his files to the dazzling visitor. Leibniz read *De analysi* and a fuller exposition of Gregory's work than that sent to him, a piece called the *Historiola* that included Newton's letter on tangents. Though he took notes on the last, he did not take notes on Newton's fluxional propositions at the end of *De analysi* nor on Gregory's method of maxima and minima. His notes concentrated on infinite series, which he saw as the subject in which British mathematics could instruct him. The absence of notes on the fluxional calculus implies that he saw nothing there he did not know already. When Collins realized the extent of his indiscretion after Leibniz's departure, he did not tell Newton what he had shown to the German mathematician. Apparently, from the content of the *Commercium epistolicum*, Newton learned only later that Leibniz had seen *De analysi*. For his part, Leibniz chose not to mention it.

Even before Leibniz's visit, Collins had been impressed enough by the reply to the *Epistola prior* to urge Newton anew to publish his method. Obsessed with the latest exchange on colors, Newton thought otherwise.

I look upon your advice as an act of singular friendship [he wrote], being I beleive censured by divers for my scattered letters in y^e *Transactions* about such things as no body els would have let come out w^thout a substantial discours. I could wish I could retract what has been done, but by that, I have learnt what's to my convenience, wch is to let what I write ly by till I am out of ye way.

Collins's expressed fear that Leibniz's method would prove more general left him wholly unmoved, and with serene confidence he described what his method could accomplish:

there is no curve line exprest by any aequation of three terms, though the unknown quantities affect one another in it, or ye indices of their dignities be surd quantities . . . but I can in less than half a quarter of an hower tell whether it may be squared or what are ye simplest figures it may be compared w^th, be those figures Conic sections or others. And then by a direct & short way (I dare say ye shortest ye nature of ye thing admits of for a general one) I can compare them. . . . This may seem a bold assertion because it's hard to say a figure may or may not be squared or compared wth another, but it's plain to me by ye fountain I draw it from . . . .

Meanwhile he completed the second response to Leibniz's questions, the *Epistola posterior*, one week after Leibniz left London for Hanover. Wickins transcribed the copy sent to London, probably the last episode in his career as Newton's amanuensis. The letter began with an auto-biographical passage on Newton's discovery of the binomial theorem and the various aborted plans for publication, a precious passage from a man not much given to self-revelation. Inexorably, the pattern of the letter drew him into *De methodis* and his fluxional method, which he discussed in tantalizing incompleteness. More even than the *Epistola prior*, the second letter was a veritable treatise on infinite series, but twice, as he approached the fluxional method, he drew back and concealed critical passages in anagrams.

Leibniz did not receive the *Epistola posterior* until the following June. As with the earlier one, Oldenburg had recognized its importance and refused to send it until he heard Leibniz was settled in Hanover and until he had a reliable carrier. On 11 June 1677, immediately upon receiving it, Leibniz penned a response filled with praise. In it, he communicated the essence of his differential calculus, asked some probing questions such as only an expert could have formulated, and virtually implored further exchange. A month later, when he had time to digest the letter, he wrote again. Oldenburg warned him in August that Newton was preoccupied with other affairs. And in September Oldenburg died. Both of Leibniz's letters were forwarded to Newton. It is impossible to imagine that he did not recognize the significance of their contents. Perhaps the long delay had aroused his suspicions, though there is no evidence to warrant a projection of later attitudes back onto 1677. The fact is that Newton had made his decision five years earlier. There is no good reason to think that he would have communicated to a German mathematician he had never met what he had been unwilling to let Collins publish five years earlier. With Oldenburg dead, he did not reply, and the correspondence lapsed.

An unpleasant paranoia pervaded the *Epistola posterior*. The auto-biographical passage insisted on the pressure of Collins and Oldenburg to publish, and he concealed two vital passages in anagrams. Two days after he sent it, he wrote to Oldenburg again: "Pray let none of my mathematical papers be printed wthout my special licence." Leibniz was probably not the object of the paranoia at this time. Rather Newton was obsessed with his correspondence on colors and allowed his frustration with it to influence his response to Leibniz. In so doing, he sowed the seeds of unlimited turmoil. In 1676, Leibniz had not published his calculus. He had not communicated it. A free and open communication from Newton would have plunged him, undeservedly, into a cruel dilemma. Before he had established any claims of his own, he would have learned that another mathematician had invented substantially the same method before him. Because the correspondence passed through Oldenburg, he would have learned it publicly. One can only speculate what the outcome would have been – and hope it would have been less discreditable to both than what did finally happen. As far as Newton is concerned, such a letter would have secured what his futile concealments gave away, an unassailable claim to prior invention of the calculus.

&a.

Meanwhile, optics had refused to leave him alone. In the autumn of 1674, Oldenburg received a letter, criticizing Newton's original paper and challenging his basic experiment, from Francis Hall (or Linus, as he latinized his name), an English Jesuit who was a professor at the English college in Liège. It inaugurated an extended exchange with Linus and his pupils which lasted into 1678 and proved to be for Newton the most trying yet. Upon the receipt of a second letter in November, Newton wrote out explicit instructions on how the experiment should be performed, cited all the others who had confirmed his description, and asked the Royal Society to try it at a meeting if they had not yet done so. Emboldened by his recollection of the spring, when he had found himself the object of admiring attention upon his first attendance at a meeting of the society, he added something more, an offer of a further paper on colors. The familiar routine had to be performed. Two-and-a-half weeks later, on 30 November, he had not yet sent the papers because, when he reviewed them, "it came into my mind to write another little scrible to accompany them." Perhaps the convenience of Wickins, who was pressed into service as amanuensis, contributed to the delay. What he finally sent on 7 December contained two items, a "Discourse of Observations," which was virtually identical to Parts I, II, and III of Book II of the *Opticks* published nearly thirty years later, and "An Hypothesis explaining the Properties of Light discoursed of in my severall Papers."

The first of the two dated from 1672, though Newton may have revised the early version into its final form in 1675. In many respects, the "Hypothesis of Light" was also not new. He had begun to draft it in 1672 as part of his reply to Hooke, and things like it had appeared already in his essay "Of Colours" in 1666. We need to read Newton's comment about it in a covering letter to Oldenburg against this background.

Sʳ. I had formerly purposed never to write any Hypothesis of light & colours, fearing it might be a means to ingage me in vain disputes: but I hope a declar'd resolution to answer nothing that looks like a controversy (unles possibly at my own time upon some other by occasion) may defend me from yᵗ fear. And therefore considering that such an Hypothesis would much illustrate yᵉ papers I promis'd to send you, & having a little time this last week to spare: I have not scrupled to describe one so far as I could on a sudden recollect my thoughts about it, not concerning my self whether it shall be thought probable or improbable so it do but render yᵉ papers I send you, and others sent formerly, more intelligible. You may see by the scratching & interlining 'twas done in hast, & I have not had time to get it transcrib'd.

For the first time, Newton undertook to reveal his thoughts about the ultimate constitution of nature; he did not find it a task that he could do lightly.

In the introduction to the "Hypothesis," as in the covering letter, Newton insisted that he sent it merely to illustrate his optical papers. He did not assume it; he did not concern himself whether the properties of light he had discovered could be explained by this hypothesis or by Hooke's or by another. "This I thought fitt to Expresse, that no man may confound this with my other discourses, or measure the certainty of one by the other, or think me oblig'd to answer objections against this script. For I desire to decline being involved in such troublesome & insignificant Disputes." It is quite impossible to reconcile the actual "Hypothesis" with Newton's deprecations of it, however. For one thing, it presented far more than an explanation of optical phenomena. For another, a feeling of intensity pervaded it. In it, Newton presented himself in his preferred role not of positive scientist but of natural philosopher confronting the entire sweep of nature. For ten years he had contemplated the order of things in solitude. Now he was disclosing, partially, to a limited audience, where ten years of speculation had carried him. No amount of feigned indifference and hard words about insignificant disputes could obscure the significance the enterprise held for him.

As far as light was concerned, the "Hypothesis" presented a largely orthodox mechanical philosophy. He assigned reflections and refractions to the causation of a universal aether which stands rarer in the pores of bodies than in free space and causes corpuscles of light to change directions by its pressure. A mechanism of vibrations in the aether explained the periodic phenomena of thin films. The "Hypothesis of Light" contained much more than an explanation of optical phenomena, however. The first half of it presented a general system of nature based on the same aether. All of the crucial phenomena that appeared in his "Quaestiones" a decade earlier appeared now in the "Hypothesis" either to be explained by aetherial mechanisms or to offer illustrative analogies. For example, the pressure of the aether explained the cohesion of bodies, and surface tension illuminated an aethereal mechanism.

Because the aether condensed continually in bodies such as the earth, there is, according to the "Hypothesis," a constant downward stream of it which impinges on gross bodies and carries them along. Newton explicitly extended this explanation of gravity to the sun and suggested that the resulting movement of aether holds the planets in closed orbits. The passage contains the first known hint of the concept of universal gravitation in Newton's papers; he did not fail to refer to it when Hooke cried plagiarism in 1686.

The "Hypothesis of Light" refuses to be presented solely as a mechanical system of nature, however. If it showed the enduring influence of the mechanical philosophy, it was an ambiguous document which contained vestiges of other influences that had begun to bear on Newton's conception of nature. One of the features that distinguished it was the prominent role of chemical phenomena, which had played no part in the "Quaestiones" ten years earlier. They epitomized the new influences that would, in the years ahead, carry him on beyond his position in 1675. I shall return to them in a different context.

Whatever his announced intention to avoid disputes, the new papers plunged Newton directly into a new round of correspondence, of explication, and very quickly of controversy. The papers were read at the Royal Society immediately, the "Hypothesis" on 9 and 16 December and, after a Christmas recess and two meetings monopolized by discussions stemming from the "Hypothesis," the "Discourse of Observations" from 20 January to 10 February. Like the paper of 1672, they caused a sensation. The Royal Society requested the immediate publication of the "Discourse," which Newton declined.

Questions also arose, most significantly from Hooke. Whether deliberately or through inadvertence, Newton had introduced him into the "Hypothesis" rather prominently, both in the introduction, which justified the whole enterprise by a reference back to Hooke's critique of 1672, and in the discussion of diffraction at the conclusion. Small wonder that Hooke rose when the reading of the "Hypothesis" was completed to assert that "the main of it was contained in his *Micrographia,* which Mr. Newton had only carried farther in some particulars."

Without much thought about the provocation he had offered, Newton erupted in anger at the charge. Because Oldenburg's letter reporting the incident does not survive, we do not know exactly what Newton heard, or for that matter exactly what happened. Indeed, none of Oldenburg's letters to Newton from this period survive, possibly a suspicious circumstance because Hooke believed that Oldenburg, with whom he was at swords' points, had deliberately fomented trouble. Even minutes of the Royal Society do not offer an independent account; Oldenburg kept them. It is not hard to believe that an incident occurred, however. Hooke was a prickly personality in his own right, and he had reason to feel aggrieved with Newton. What the "Hypothesis" poured on his wounds was more embalming fluid than balm. For his pains he now received a further dose of gall. Hooke's hypothesis of light, Newton asserted, was merely an embroidery on Descartes's. His own was entirely different, to the extent that the experiments, which were new to Hooke, on which Newton based his treatment of thin films undermined everything Hooke had said about the subject. The more he wrote, the hotter Newton became. True, he had learned about colors in thin films from Hooke. Hooke had confessed that he did not know how to measure the thickness of the films, however, "& therefore seing I was left to measure it my self I suppose he will allow me to make use of what I tooke ye pains to find out." Three weeks of thinking about it left Newton even more incensed. Initially, he was inclined to grant that he got the idea of vibrations in the aether from Hooke. Now he retracted that as well; it was a common idea. "I desire Mr Hooke to shew me therefore, I say not only ye summ of ye Hypothesis I wrote, wch is his insinuation, but any part of it taken out of his *Micrographia:* but then I expect too that he instance in what's his own."

The handling of Newton's first letter tends to confirm Hooke's suspicions that Oldenburg egged him on. Though Oldenburg read a passage from it to the Royal Society on 30 December, he neither read the comment on Hooke nor told Hooke about it. Hooke heard it with surprise at the meeting on 20 January. At that point, he took matters into his own hands and wrote directly to Newton the same day. He feared Newton had been misinformed about him, a "sinister practice" which had been used against him before. He protested his disapproval of contention, his desire to embrace truth by whomever discovered, and the value he placed on Newton's "excellent Disquisitions," which proceeded farther than any thing he had done. Finally, he proposed a correspondence in which the two could discuss philosophical matters privately. "This way of contending I believe to be the more philosophicall of the two, for though I confess the collision of two hard-to-yield contenders may produce light yet if they be put together by the ears of other's hands and incentives, it will produce rather ill concomitant heat which serves for no other use but . . . kindle cole [*sic*]."

Newton replied in kind, calling Hooke "a true Philosophical spirit." "There is nothing wch I desire to avoyde in matters of Philosophy more then contention," he agreed, "nor any kind of contention more then one in print . . . ." Accepting the offer of a private correspondence, he went on to praise Hooke's contribution to optics. "What Des-Cartes did was a good step. You have added much several ways, & especially in taking ye colours of thin plates into philosophical consideration. If I have seen further it is by standing on ye sholders of Giants." Sentiments too lofty drift away from human reality. A lack of warmth was evident on both sides. Neither man endeavored to institute the philosophic correspondence both professed to want, and their basic antagonism remained undissolved.

❧

Another correspondence refused to go away, the one instituted by Linus, which provided the occasion for the two papers of December. That same month, a letter from Liège written by John Gascoines, a pupil of Linus, informed Oldenburg and Newton that Linus was dead but that Gascoines intended to defend his professor's honor. In June, at the time of the *Epistola prior*, a new letter arrived from a third correspondent, Anthony Lucas, another English Jesuit, whom Gascoines had recruited to take up cudgels too heavy for him. Lucas began by conceding the sole point that hitherto had been in contention: A prism does project a spectrum elongated perpendicularly to the axis of the prism, though his spectrum did not have the same proportions as Newton's. Lucas proceeded to relate the results of nine other experiments he had performed to test Newton's theory. Far from confirming it, their outcomes seemed to negate it. Through four years of discussion, Newton had challenged opponents to bring forward experiments instead of hypotheses. He greeted Lucas's experiments with the contrary of a reasoned response, however. As the correspondence continued, he became increasingly agitated and irrational. He convinced himself that the Liègois (papists, of course) had formed a conspiracy to engage him in perpetual disputation and to undermine his credit. He refused to discuss Lucas's experiments but insisted that Lucas discuss his.

Tis ye truth of my experiments which is ye business in hand [he stormed]. On this my Theory depends, & which is of more consequence, ye credit of my being wary, accurate and faithfull in ye reports I have made . . . .

"I see I have made my self a slave to Philosophy," he exclaimed to Oldenburg in desperation, "but if I get free of Mr Linus's buisiness I will resolutely bid adew to it eternally, excepting what I do for my privat satisfaction or leave to come out after me. For I see a man must either resolve to put out nothing new or to become a slave to defend it." Recall that at the time he wrote, Newton's "slavery" consisted of five replies to Liège, totaling fourteen printed pages, over a period of a year. Recall also that he had completed the *Epistola posterior* less than a month before. When a third letter from Lucas arrived in February 1677, Newton decided on a different mode of answer and began to plan a volume on optics that would include his papers and the correspondence they had provoked. Then fire struck his chamber, destroying part of the collection of papers. Though he tried briefly to get new copies, he finally abandoned the project.

Fourteen years later, Abraham de la Pryme, a student in Johns, recorded in his diary a story he had heard.

Febr: [1692] What I heard to-day I must relate. There is one Mr. Newton . . . fellow of Trinity College, that is mighty famous for his learning, being a most excellent mathematician, philosopher, divine, etc. . . . but of all the books that he ever writt there was one of colours and light, established upon thousands of experiments, which he had been twenty years of making, and which had cost him many a hundred of pounds. This book which he valued so much, and which was

so much talk'd off, had the ill luck to perish and be utterly lost just when the
learned author was almost at putting a conclusion at the same, after this manner.
In a winter morning, leaving it amongst his other papers on his studdy table, whilst
he went to chappel, the candle which he had unfortunately left burning there too
cachd hold by some means or other of some other papers, and they fired the
aforesayd book, and utterly consumed it and several other valuable writings, and
that which is most wonderful did no further mischief. But when Mr. Newton
came from chappel and had seen what was done, every one thought he would have
run mad, he was so troubled thereat that he was not himself for a month after. A
large account of this his system of light and colours you may find in the transactions of the Royal Society, which he had sent up to them long before this sad
mischance happened to them.

De la Pryme's story is usually attached to Newton's established breakdown in the autumn of 1693, which a story Huygens heard, also involving
a fire, tends to support. However, the date of de la Pryme's entry ante-
dates the breakdown of 1693 by more than eighteen months, and the past
perfect tense in the final sentence does not seem to place the fire in the
recent past. The entry could refer to the fire, authenticated by a body of
other evidence, that halted an optical publication in the winter of 1677–8.
There is a hiatus in Newton's correspondence from 18 December to
February, though his correspondence in this period was very light in any
case. Twice at least in earlier correspondence, with Hooke and with
Huygens, he had lost partial control of himself as the heat of his own
vehemence swept him away, and the tone of his letters to Lucas implies a
complete loss of control that is compatible with a breakdown. As later in
1693, Newton was in a state of acute intellectual tension throughout the
1670s, not just from answering objections to his optics but even more
from other studies then foremost in his mind, studies which excited him
keenly. Another parallel with 1693 may also be relevant. The crisis in his
relations with Fatio de Duillier at that time had its counterpart in Wickins's decision to leave Trinity.

When he discarded the intended publication, Newton wrote two fur-
ther letters to Lucas on the same day, 5 March 1678, one answering
Lucas's first two and the other answering his third (of February 1677).
Even the earlier letters, furious as they were, could not have prepared
Lucas for the flood of paranoia that now burst over him.

Do men use to press one another into Disputes? Or am I bound to satisfy you? It
seems you thought it not enough to propound Objections unless you might insult

over me for my inability to answer them all, or durst not trust your own judgement in choosing ye best. But how know you y^t I did not think them too weak to require an answer & only to gratify your importunity complied to answer one or two of ye best? How know you but y^t other prudential reasons might make me averse from contending wth you? But I forbeare to explain these things further for I do not think this a fit Subject to dispute about, & therefore have given these hints only in a private Letter. of wch kind you are also to esteem my former answer to your second. I hope you will consider how little I desire to explain your proceedings in public & make this use of it to deal candidly wth me for ye future.

Arrogant and brutal, the two letters made it clear that nothing less than the public humiliation of his antagonists could satisfy Newton – hateful letters, did not our knowledge of the circumstances incline us to sympathy with the author's anguish.

The correspondence gave one last spasm before it died. In May Newton acknowledged a letter from Lucas, though he probably did not answer it. Later that month, he heard that there was another waiting for him in London.

Mr Aubrey

I understand you have a letter from Mr Lucas for me. Pray forbear to send me anything more of that nature.

With that he brought his correspondence about colors to an end. Oldenburg was dead. Newton had ceased to correspond with Collins. As far as he could, he isolated himself. To the best of our knowledge he wrote only two letters, one (which has not survived) to Arthur Storer and one to Robert Boyle, between June 1678 and December 1679. To the end of his days Newton remembered this withdrawal as a conscious decision and considered that it had marked an epoch in his life. "Its now about fifty years," he wrote to Mencke in 1724, "since I began for the sake of a quiet life to decline correspondencies by Letters about Mathematical & Philosophical matters finding them tend to disputes and controversies . . . ."

6

Rebellion

NEWTON'S REPEATED PROTESTATION that he was engaged in other studies supplied an ever-present theme to his correspondence of the 1670s. Already in July 1672, only six months after the Royal Society discovered him to be a man supremely skilled in optics, he wrote to Oldenburg that he doubted he would make further trials with telescopes, "being desirous to prosecute some other subjects." Three-and-a-half years later, he put off the composition of a general treatise on colors because of unspecified obligations and some "buisines of my own wch at present almost take up my time & thoughts." Apparently the other business was not mathematics, because later in 1676 he hoped the second letter for Leibniz would be the last. "For having other things in my head, it proves an unwelcome interruption to me to be at this time put upon considering these things." He was not only preoccupied, he was almost frantic in his impatience. "S^r," he concluded the letter, "I am in great hast, Yours . . . ." In great haste because of what? Surely not because of ten lectures on algebra that he purportedly delivered in 1676. And not because of pupils or collegial duties, for he had none of either. Only the pursuit of Truth could so drive Newton to distraction that he resented the interruption a letter offered. Newton was in a state of ecstasy again. If mathematics and optics had lost the capacity to dominate him, it was because other studies had supplanted them.

One of the studies was chemistry. Collins mentioned his absorption in it twice in letters to Gregory. Years later, when he chatted with Conduitt about his early life in Cambridge, Newton himself mentioned that Wickins helped in his "chymical experiments." His interest in it developed somewhat later than his interest in natural philosophy. When he composed the "Quaestiones quaedam philosophicae" in the mid-1660s, he entered almost nothing that one would call chemistry, even though Robert Boyle was one of the major sources of his new mechanical philosophy. When he extended his notes on a number of the headings under "Quaestiones" in a new notebook, however, chemistry did begin to appear, and the notes indicate that Boyle supplied his introduction to the subject. Newton's ability to organize what he learned so that he could retrieve it was a significant aspect of his genius. Years later, in a paper he prepared at the Mint, he described a process to refine gold and silver with lead that he noted at this time, and he used some of the same language he had entered fifty years earlier.

Not all of the entries in a chemical glossary he composed at this time confined themselves to straightforward, prosaic chemistry – or "rational chemistry," as those call it who wish to pretend that Newton did not leave behind a vast collection of alchemical manuscripts. He included quite a few entries on mercury, including mercury sublimate, which "opens" copper, tin, and silver, but not gold. "Yet perhaps," he added, "there may bee Sublimates made (as by subliming common sublimate & Sal Armoniack well powdered together) wch besides notable operations on other metalls, may act upon Gold too." One entry described Boyle's *menstruum peracutum*, which dissolved gold and even carried some gold with it in distillation. Boyle invested the *menstruum peracutum* with alchemical significance; Newton's entry implied that he did too. Antimony and its power to purify gold appeared. As with the refining of gold by lead, Newton later employed his knowledge of refining by antimony in an emotionally charged memorandum when the standard of his coinage was impugned at the trial of the pyx, a procedure to ensure quality at the Mint, in 1710. His early glossary also included instructions to make regulus of antimony, regulus of Mars, and "Regulus Martis Stellatus" the star regulus of Mars, which would soon figure prominently in an explicitly alchemical setting.

In similar fashion, the chemical notebook changed its character. Notes from George Starkey's *Pyrotechny Asserted* succeeded those from Boyle. Starkey was the pseudonymous Eirenaeus Philalethes, whose numerous treatises on alchemy exercised enormous influence on Newton. One of the last sections of the notebook, added perhaps a decade after the initial set, carried the heading, "Of ye work wth common ⊙ [gold]." He drew the content of the entry from Philalethes' commentary on Ripley.

No solid evidence allows us to date with precision Newton's plunge into alchemy. A number of items suggest 1669. The completion of his optical research before his appointment to the Lucasian chair may have cleared the way for a new intellectual passion. His impatience with questions about the theory of colors in the 1670s sprang in part from his total absorption in a new investigation.

The order of development of Newton's chemical notebook was significant. He did not stumble into alchemy, discover its absurdity, and make his way to sober, "rational," chemistry. Rather he started with sober chemistry and gave it up rather quickly for what he took to be the greater profundity of alchemy. The latest notes attributed to Boyle referred to his *Essay of . . . Effluviums* of 1673. An unattributed recipe for making phosphorus (which began with the heroic instruction "Take of Urin one Barrel") undoubtedly stemmed from Boyle's investigation of phosphorus in the early 1680s, but an isolated recipe for a new and unusual substance is a different matter from notes on sustained reading. Boyle himself was deeply involved in alchemy in any case, and once they became acquainted, the two men corresponded on the subject until Boyle's death in 1691. Meanwhile, reading that began with Boyle in the 1660s turned overwhelmingly to explicitly alchemical authors about 1669. His accounts show that on his trip to London that year he purchased the great collection of alchemical writings *Theatrum chemicum* in six heavy quarto volumes. He also purchased two furnaces, glass equipment, and chemicals. Possibly some practitioner of the Art introduced Newton to it. Evidence exists that Cambridge had its adepts while Newton was there. We are not obliged to look for an alchemical father, however. Newton had already found his way alone to a number of studies. With collections such as *Theatrum chemicum* available, his independent discovery of alchemy would have been easy enough.

Solid evidence further shows that however it began, Newton's alchemical activity included his personal introduction into the largely clandestine society of English alchemists. His reading in alchemy was not confined to the printed word. Among his manuscripts is a thick sheaf of alchemical treatises, most of them unpublished, written in at least four different hands. Because Newton copied out five of the treatises plus some recipes, the collection appears to have been loaned to him for study but then, for whatever reason, not returned. In the late 1660s, he copied

Philalethes' "Exposition upon Sir George Ripley's Epistle to King Edward IV" from a version which differed from published ones, though it agreed with two manuscripts now in the British Library. He took extensive notes from a manuscript of Philalethes' "Ripley Reviv'd" about ten years before it was published. During the following twenty-five years, Newton continued to receive a flow of alchemical manuscripts which he himself copied.

These manuscripts offer one of the most intriguing aspects of his career in alchemy. Where did they come from? The Philalethes manuscripts circulated initially among the group of alchemists associated with Samuel Hartlib in London. Hartlib had died well before Newton took up alchemy, but he may have been in touch with remnants of the group. Because William Cooper, who kept a shop at the sign of the Pelican in Little Britain, later published "Ripley Reviv'd" and at least two other treatises that Newton copied, the contact may have been through him. Robert Boyle had known the Hartlib circle as well as Philalethes–Starkey, though it seems clear that Newton first met Boyle in 1675. One of the copied manuscripts concluded with letters dated 1673 and 1674 from A. C. Faber to Dr. John Twisden with Twisden's notes on them and on the manuscript. Faber (A. D. rather than A. C.) was a physician to Charles II who published a treatise on potable gold. Twisden, also a physician in London known for his defense of Galenic medicine, does not seem a likely clandestine alchemist, but the notes attributed to him are those of a serious practitioner. There was at least the possibility of personal contact and direct transmission in connection with this paper. On another manuscript, "Manna," which is not in his hand, Newton entered two pages of notes and variant readings "collected out of a M.S. communicated to Mr F. by W. S. 1670, & by Mr F. to me 1675." Professor B.J.T. Dobbs has argued plausibly that "Mr F." was Ezekiel Foxcroft, a fellow of King's who died in that same year of 1675. Foxcroft, the nephew of Benjamin Whichcote, relative by marriage of John Worthington, and friend of Henry More (all Cambridge Platonists), translated the Rosicrucian tract "The Chymical Wedding," which was published fifteen years after his death. Newton read it and made notes on it at that time. Whether or not "Mr F." was Ezekiel Foxcroft, the essential mystery of the alchemical manuscripts remains unclarified. The man who isolated himself from his colleagues in Trinity and discouraged correspondence from philosophical peers in

London apparently remained in touch with alchemists from whom he received manuscripts.

The mystery refuses to be ignored. The manuscripts survive – unpublished alchemical treatises, copied by Newton, the originals of which are unknown. The not very illuminating references to Twisden, Faber, "W. S.," and "Mr F." excepted, the network of acquaintance that brought them to him left virtually no tangible evidence behind. In March 1683, one Fran. Meheux wrote to Newton from London about the success of a third alchemist, identified only as "hee," in extracting three earths from the first water. Meheux's letter mentioned a continuing correspondence, but the letters have disappeared. Meheux and "hee" have all the substance of shadows. In 1696, an unnamed and equally shadowy figure, a Londoner acquainted with Boyle and Edmund Dickinson (a well-known alchemist whom Charles II had patronized), visited Newton in Cambridge to discourse on alchemy. They did not meet by chance; the man came to find him. Newton recorded the conversation in a memorandum. Alchemy formed the initial subject of a correspondence with Robert Boyle which commenced in 1676. His friendships with John Locke and Fatio de Duillier involved alchemy, but both of them began only in the late 1680s. Otherwise nothing. One of the major passions of his life, as testified by a vast body of papers which stretched over thirty years, a pursuit which included contact with alchemical circles as attested by his copies of unpublished treatises, remained largely hidden from public view and remains so today.

Newton's own manuscripts establish the fact that about 1669 he began to read extensively in alchemical literature. His notes on the reading survive, the hand not datable with precision but unmistakably from the general period of the late 1660s and perhaps 1670–1. In her recent study of Newton's early alchemy, Professor Dobbs asserts that Newton probed "the whole vast literature of the older [i.e., pre–seventeenth century] alchemy as it has never been probed before or since." He also studied seventeenth-century alchemists, especially Sendivogius, d'Espagnet, and Eirenaeus Philalethes, with equal intensity. Much of Newton's attention to alchemy came later. I have carried out a rather careful quantitative study of the alchemical manuscripts he left behind, in which I divide them into three chronological groups. Of the total, which I estimate to include well over a million words devoted to alchemy, about one-sixth appear to stem from the period before 1675. In his usual fashion, Newton purchased a notebook in which he entered twelve general headings and a number of subheadings under which to organize the fruits of his reading – headings such as "Conjunctio et liquefactio," "Regimen per ascensum in Caelum & descensum in terram," and "Multiplicatio." In this case, he did not carry the plan out beyond a small number of entries. His later Index chemicus would supply its lack on a heroic scale. Meanwhile, the reading proceeded apace.

Whatever else alchemy meant to him, Newton was always convinced that the treatises he read referred to changes that material substances undergo. It was his goal to penetrate the jungle of luxuriant imagery in order to find the process common to all the great expositions of the Art. To assert as much is not to say that the chemistry he pursued would have been acceptable in the scientific academies of his day, or that scientists of the twentieth century would be willing even to recognize it as chemistry. Nevertheless, he did understand that chemical processes, not mystical experience expressed in the idiom of chemical processes, formed the content of the Art. Thus his reading in the literature of alchemy proceeded hand in hand with laboratory experimentation. The progress made of late in penetrating the maze of his alchemical endeavor has rested on the correlation of his surviving experimental notes with the alchemical manuscripts.

Most of the experimental notes derive from 1678 and after. There are some undated ones in his chemical notebook, however, which seem definitely to have belonged to the late 1660s and early 1670s. His earliest experiments, based on Boyle and showing perhaps the influence of Michael Maier as well, attempted to extract the mercury from various metals. In the intellectual world of alchemy, mercury – not common quicksilver, but the mercury of the philosophers – was the common first matter from which all metals were formed. Liberating it from its fixed form in metals, cleansing it of contaminating feces, was equivalent to vivifying it and making it fit for the Work. The two images found here, images of purification and of vivification, which included generation by male and female, pervaded the alchemical literature that Newton read. His laboratory notes reveal his attempts to extract the mercury of the philosophers by different means, as well as his experimentation with more powerful alchemical methods involving the star regulus of Mars, that is the regulus of antimony made with iron.

჈

From the testimony of others, from his own reading notes, and from his experiments with substances alchemically significant, it is clear beyond doubt that Newton devoted great attention to alchemy in the late 1660s and early 1670s. We are left to decide for ourselves what his purpose may have been. As everyone knows, alchemy sought to make gold. Nothing whatever in the vast corpus of Newton's alchemical manuscripts even hints that gold-making, in the vulgar sense of the phrase, ever dominated Newton's concern. While Newton was not indifferent to his material welfare, it was never money that kept him from his meals and drove him to distraction. Truth and Truth alone held that power over him. To the great figures and monuments of the alchemical tradition, the men and works that Newton studied, Truth was also the goal of the Art. As Elias Ashmole insisted in the preface to his *Theatrum chemicum britannicum*, gold-making was the lowest use to which the adepts applied their knowledge.

For they being lovers of Wisdome more then Worldly Wealth, drove at higher and more Excellent Operations: And certainly He to whom the whole Course of Nature lyes open, rejoyceth not so much that he can make Gold and Silver, or the Divells to become Subject to him, as that he sees the Heavens open, the Angells of God Ascending and Descending, and that his own Name is fairely written in the Book of life.

The philosophical tradition of alchemy had always regarded its knowledge as the secret possession of a select few who were set off from the vulgar herd both by their wisdom and by the purity of their hearts. By the time he turned seriously to alchemy, Newton had carried two investigations of capital importance to completion; he could not have doubted his right to claim membership in an intellectual elite. We know less of what he thought about the purity of his heart, but convictions on that score are endemic throughout humankind.

The concept of a secret knowledge for a select few aside, all the foregoing characteristics applied as well to the mechanical philosophy, which Newton had recently embraced. In the nature of the truth they offered, however, the two philosophies differed profoundly. In the mechanical philosophy, Newton had found an approach to nature which radically separated body and spirit, eliminated spirit from the operations of nature, and explained those operations solely by the mechanical necessity of particles of matter in motion. Alchemy, in contrast, offered the quintessential embodiment of all the mechanical philosophy rejected. It looked upon nature as life instead of machine, explained phenomena by the activating agency of spirit, and claimed that all things are generated by the copulation of male and female principles. Among his "Notable Opinions" that he collected some ten years later Newton included the argument of Effararius the Monk that the stone is composed of body, soul, and spirit, that is, imperfect body, ferment, and water.

For a heavy and dead body is imperfect body per se. The Spirit that purges, lightens, and purifies body is water. The soul that gives life to imperfect body when it does not have it, or raises it to a higher plane, is ferment. Body is Venus and feminine; spirit is Mercury and masculine; soul is the Sun and the Moon.

And in a later collection of "Enlightening Opinions and Notable Conclusions," Newton included an unattributed expression of the concept of sexual generation to which Effararius alluded in his final sentence.

A double mercury is the sole first and proximate matter of all metals, and these two mercuries are the masculine and feminine semens, sulfur and mercury, fixed and volatile, the Serpents around the caduceus, the Dragons of Flammel. Nothing is produced from masculine or feminine semen alone. For generation and for the first matter the two must be joined.

Newton also met in alchemy another idea that refused to be reconciled with the mechanical philosophy. Where that philosophy insisted on the inertness of matter, such that mechanical necessity alone determines its motion, alchemy asserted the existence of active principles in matter as the primary agents of natural phenomena. Especially it asserted the existence of one active agent, the philosophers' stone, the object of the Art. Images of every sort were applied to the stone, all expressing a concept of activity utterly at odds with the inertness of mechanical matter characterized by extension alone. Flammel called it "a most puissant & invincible king"; Philalethes, the "miracle of the world" and "the subject of wonders." The author of *Elucidarius* insisted that "it is impossible to express [its] infinite virtues . . . ." In Sendivogius and Philalethes, the activity sometimes took on the specific form of an attraction, and they called it a magnet. Philosophic reformers such as Descartes had explicitly worked to eliminate "occult" concepts, such as attractions, from natural philosophy; they had invented whirlpools of various invisible particles to explain away the apparent fact of magnetism. Not Sendivogius and Philalethes. To them, the magnet offered an image of the operation of nature. "They call lead a magnet," Newton recorded in his early notes on Sendivogius, "because its mercury attracts the seed of Antimony as the magnet attracts the Chalybs." And again he noted that "our water" is drawn out of lead "by the force of our Chalybs which is found in the belly of Ares." In a note, Newton explained that this meant "the force of our sulfur which lies hidden in Antimony."

It is necessary, I believe, to see Newton's interest in alchemy as a manifestation of rebellion against the confining limits that mechanistic thought imposed on natural philosophy. If the pursuit of Truth expressed the essence of his life, there is no reason to expect that he should have remained satisfied forever with his first love. Mechanical philosophy had surrendered to his desire, perhaps too readily. Unfulfilled, he continued the quest and found in alchemy, and in allied philosophies, a new mistress of infinite variety who never seemed fully to yield. Where others cloyed she only whet the appetite she fed. Newton wooed her in earnest for thirty years.

Perhaps *rebellion* is too strong a word, and I should speak rather of partial rebellion. Newton never wholly abandoned his first love. He never ceased to be a mechanical philosopher in a fundamental and important sense. He always believed that particles of matter in motion constitute physical reality. Where mechanical philosophers of strict persuasion insisted that particles of matter in motion alone constitute physical reality, however, Newton came rather early to find those categories too confining to express the reality of nature. The significance of alchemy in his intellectual odyssey lay in the broader vistas it opened to him, additional categories to supplement and complete the narrow mechanistic ones. His enduring fame derived from his seizing the possibilities thus spread before him.

The conviction found in Newton's alchemical papers of the early 1670s, that mechanical science had to be completed by a more profound natural philosophy which probed the active principles behind particles in motion, repeated itself in the "Hypothesis of Light" of 1675, though he disguised it considerably, perhaps because of the audience. On the surface, the "Hypothesis" presented a mechanical cosmology based on a universal aether, and for three hundred years it has been read as a representative expression of seventeenth-century mechanical philosophy. It contained strange elements, however, though they look less strange after one has read the "Vegetation of Metals," one of his early alchemical papers. Several times the "Hypothesis" referred to a "secret principle of unsociablenes" by which fluids and spirits do not mix with some things but do with others. Active principles appeared as well. He imagined the condensation of the aether in fermenting and burning bodies, and its exhalation in vapors, so that the whole earth "may be every where to the very center in perpetuall working."

For nature is a perpetuall circulatory worker, generating fluids out of solids, and solids out of fluids, fixed things out of volatile, & volatile out of fixed, subtile out of gross, & gross out of subtile, Some things to ascend & make the upper terrestriall juices, Rivers and the Atmosphere; & by consequence others to descend for a Requitall to the former.

Small wonder that the "Hypothesis" has recently been called an alchemical cosmology.

Not long after he composed the "Hypothesis," Newton read in the *Philosophical Transactions* an account by "B.R." of a special mercury that heated gold when mixed with it. B.R. asked for advice as to whether he should publish the recipe for the mercury. As far as we know, Newton was the only one who offered an opinion to Robert Boyle, whom Newton correctly understood to be B.R. The most interesting thing about the letter to Boyle is the fact that he wrote it. At the very time when he was frantically trying to terminate his correspondence on optics and mathematics, he volunteered a letter on alchemy which looks like an effort to initiate a correspondence. Later evidence confirms that one ensued, the only sustained one we know of during his years of silence, and for that matter, after the initial letter, a direct correspondence without intermediary.

≈

If, as the manuscript evidence suggests, Newton's active involvement in alchemy slackened for a time after the early years of the decade, he did not lack for other interests. His papers show that about this time he turned to a new field of study, theology. Perhaps it is wrong to call it "new." Speculation about his reading in his stepfather's library aside, there is solid evidence of early theological interest. Four of the ten books known by his accounts and by his dated signature to have been purchased soon after his arrival in Cambridge were theological. Nevertheless, no body of theological manuscripts survives from earlier than about 1672. At that time, Newton was completing his fourth year as a Master of Arts and fellow of Trinity. Within the next three years, he would need to be ordained to the Anglican clergy or face expulsion from the college. The beginning of serious theological study may have stemmed from the approaching deadline. Whatever the cause, the fact itself cannot be denied. Nor should we imagine that he devoted himself to theology reluctantly, for the subject quickly seized him as others had before. His notes reveal a massive commitment to it. There are very few secure dates internal to the manuscripts, and in locating them chronologically, one is thrown back primarily on the uncertain evidence of handwriting. Nevertheless, there can be no reasonable question that at least part of the time, when Newton expressed impatience at the interruptions caused by optical and mathematical correspondence during the 1670s, it was theology that preoccupied him. He jotted down a number of theological references on a draft of his letter of 4 December 1674, in which he told Oldenburg that he intended "to concern my self no further about promotion of Philosophy."

If it is impossible to date most of the manuscripts with precision, so also it is impossible to be certain of their order. Surely the standard Newtonian exercise in organization was among the earliest, however. In a notebook he entered a number of headings that summarized Christian theology: "Attributa Dei," "Deus Pater," "Deus Filius," "Incarnatio," "Christi Satisfactio, & Redemptio," "Spiritus Sanctus Deus," and the like. Apparently he intended to use the notebook to systematize his study of the Bible – the references that he entered, the foundation of his extensive knowledge of the holy writings, came almost entirely from the Scriptures. Although the list of headings appears unexceptionably orthodox, Newton's entries under them suggest that certain doctrines, which had the inherent capacity to draw him away from orthodoxy, had begun to fascinate him. In his original list, he alloted one folio to "Christi Vita" and the following one to "Christi Miraculi." When an earlier entry spilled over onto the first, he joined it to the second, and he entered nothing at all under the combined heading. He left five full folios, or ten pages, for the heading "Christi Passio, Descensus, et Resurrectio" and two folios, or four pages, for "Christi Satisfactio, & Redemptio." He filled fewer than two of the ten pages intended for the first and under one of the four pages intended for the second. The heading that had spilled over its allotted two pages was "Deus Filius." Under it, Newton collected passages from the Bible that defined the relation of the Son to God the Father. From Hebrews 1 he quoted verses 8–9, which say that God set Christ on his right hand, called him God, and told him that because he had loved righteousness "therefore God, even <u>thy God</u>, hath annointed thee with the oil of gladness above thy fellows." Opposite the two words he had underlined, Newton inserted a marginal note: "Therefore the Father is God of the Son [when the Son is considered] as God." A later entry reinforced the implication of the note.

Concerning the subordination of Christ see Acts 2.33.36. Phil 2.9.10. 1 Pet 1.21. John 12.44. Rom 1.8 & 16. 27. Acts 10.38 & 2.22. 1 Cor 3.23, & 15.24, 28. & 11.3. 2 Cor. 22, 23.

Under "Deus Pater," he had already entered half a page of references on the same topic, including three that began to sound rather pointed.

There is one God & one Mediator between God & Man ye Man Christ Jesus. 1 Tim. 2.5.

The head of every man is Christ, & ye head of ye woman is ye man, & the head of Christ is God. 1 Cor. 11.3.

He shall be great and shall be called ye son of y^e <u>most high</u>. Luke 1.32.

It was Newton who underlined "most high." The reiterated implication in the two headings of a real distinction between God the Father and God the Son suggests that almost the first fruit of Newton's theological study was doubt about the status of Christ and the doctrine of the Trinity. If the approaching need for ordination had started Newton's theological reading, the reading itself started to threaten ordination.

In the other end of the notebook, Newton entered a new set of headings under which he recorded notes from other theological readings, mostly from early fathers of the church. The nature of the headings (e.g., "De Trinitate," "De Athanasio," "De Arrianis et Eunomianis et Macedonianis," "De Haerisibus et Haereticis"), together with a couple of citations of fathers at the end of headings in the other end, strongly implies that this end of the notebook involved new reading undertaken to explore the questions already raised. The content of the notes exercised lasting influence over Newton's life. He drew up an index to facilitate access to them, and a number of entries in later hands demonstrate that he did return to them. The convictions that solidified as he collected the notes remained unaltered until his death.

The longest entry, "De Trinitate" (Concerning the Trinity), filled nine pages. The passage was studious rather than contentious. Newton returned to the works of the men who had formulated trinitarianism – Athanasius, Gregory Nazianzen, Jerome, Augustine, and others – to inform himself correctly about the doctrine. Other "Observations upon Athanasius's works," together with notes elsewhere, contributed to the same goal. More than the doctrine interested him. He became fascinated with the man Athanasius and with the history of the church in the fourth century, when a passionate and bloody conflict raged between Athanasius and his followers, the founders of what became Christian orthodoxy, on the one hand, and Arius and his followers, who denied the trinity and the status of Christ in the Godhead, on the other; and he read extensively about them. Indeed, once started, Newton set himself the task of mastering the whole corpus of patristic literature. In addition to those mentioned previously, Newton cited in the notebook Irenaeus, Tertullian, Cyprian, Eusebius, Eutychius, Sulpitius Severus, Clement, Origen, Basil, John Chrysostom, Alexander of Alexandria, Epiphanius, Hilary, Theodoret, Gregory of Nyssa, Cyril of Alexandria, Leo I, Victorinus Afer, Rufinus, Manentius, Prudentius, and others. He seemed to know all the works of prolific theologians such as Augustine, Athanasius, and Origen. There was no single patristic writer of importance whose works he did not devour. And always, his eye was on the allied problems of the nature of Christ and the nature of God.

The conviction began to possess him that a massive fraud, which began in the fourth and fifth centuries, had perverted the legacy of the early church. Central to the fraud were the Scriptures, which Newton began to believe had been corrupted to support trinitarianism. It is impossible to say exactly when the conviction fastened upon him. The original notes themselves testify to early doubts. Far from silencing the doubts, he let them possess him. "For there are three that bear record in heaven, the Father, the Word, and the Holy Ghost: and these three are one." Such is the wording of 1 John 5:7, which he read in his Bible. "It is not read thus in the Syrian Bible," Newton discovered. "Not by Ignatius, Justin, Irenaeus, Tertull. Origen, Athanas. Nazianzen Didym Chrysostom, Hilarius, Augustine, Beda, and others. Perhaps Jerome is the first who reads it thus." "And without controversy great is the mystery of godliness: God was manifest in the flesh . . . ." Thus 1 Timothy 3:16, in the orthodox version. The word *God* is obviously critical to the usefulness of the verse to support trinitarianism. Newton found that early versions did not contain the word but read only, "great is the mystery of godliness which was manifested in the flesh." "Furthermore in the fourth and fifth centuries," he noted, "this place was not cited against the Arians."

The corruptions of Scripture came relatively late. The earlier corruption of doctrine, which called for the corruption of Scripture to support it, occurred in the fourth century, when the triumph of Athanasius over Arius imposed the false doctrine of the trinity on Christianity. Central to trinitarianism was the adjective *homoousios*, which was used to assert that the Son is consubstantial (*homoousios*) with the Father. Newton tended to call the Athanasians "homousians." In an early sketch of the history of the church in the fourth century, he described how the opponents of Arius in the Council of Nicaea wanted to base their argument solely on scriptural citations as they rejected Arianism and affirmed their own convictions that the Son is the eternal uncreated *logos*. However, the debate drove them to assert that the Son is *homoousios* with the Father even though that word is not in Scripture. "That is, when the Fathers were not able to assert the position of Alexander [the Bishop of Alexandria, who had charged Arius with heresy] from the scriptures, they preferred to desert the scriptures than not to condemn Arius." Eusebius of Nicomedia had introduced the word *homoousios* into the debate as a clearly heretical, intolerable consequence of the anti-Arian position.

Thus you see these fathers took ye word not from tradition but from Eusebius's letter, in wch though he urged it as a consequence from Alexander's doctrin which he thought so far from ye sense of ye Church y^t even they themselves would not admit of it, yet they chose it for it's being opposite to Arius.

Athanasius claimed that the orthodox use of the term *homoousios* did not begin with the Council of Nicaea but could be found, for example, in the writing of the third-century father Dionysius of Alexandria. Careful study revealed to Newton that Athanasius had deliberately distorted Dionysius to make it appear that he accepted a term which in fact he considered heretical. Other early fathers had been tampered with as well. Words were "foisted in" the epistles of Ignatius, of the second century, for example, to give them a trinitarian flavor. Athanasius had also distorted the proclamation of the Council of Serdica for the same purpose.

In Newton's eyes, worshipping Christ as God was idolatry, to him the fundamental sin. "Idolatria" had appeared among the original list of headings in his theological notebook. The special horror of the perversion that triumphed in the fourth century was the reversion of Christianity to idolatry after the early church had established proper worship of the one true God. "If there be no transubstantiation," he wrote in the early 1670s, "never was Pagan Idolatry so bad as the Roman, as even Jesuits sometimes confess." Newton held that the pope in Rome had aided and abetted Athanasius and that the idolatrous Roman church was the direct product of Athanasius' corruption of doctrine.

In the end – and the end did not wait long – Newton convinced himself that a universal corruption of Christianity had followed the central corruption of doctrine. Concentration of ecclesiastical power in the hands of the hierarchy had replaced the polity of the early church. The perverse institution of monasticism sprang from the same source. Athanasius had patronized Anthony, and the "homousians" had introduced monks into ecclesiastical government. In the fourth century, trinitarianism fouled every element of Christianity. Though he did not say so, he obviously believed that the Protestant Reformation had not touched the seat of infection. In Cambridge of the 1670s his was strong meat indeed. It is not hard to understand why Newton became impatient with interruptions from minor diversions such as optics and mathematics. He had committed himself to a reinterpretation of the tradition central to the whole of European civilization. Well before 1675, Newton had become an Arian in the original sense of the term. He recognized Christ as a divine mediator between God and humankind, who was subordinate to the Father Who created him.

His new convictions probably influenced his relation to Cambridge. Whatever the factors in his personality and position that made for isolation, his heretical convictions in a society of pliant orthodoxy operated far more powerfully to the same end. Cambridge was tolerance itself in

regard to performance; it did not extend tolerance to belief. Because any
discussion was fraught with the danger of ruin, Newton chose silence.
Significantly, with one exception, none of his theological papers appear in
the hand of Wickins. The one exception, an anti-Catholic interpretation
of Revelation that Wickins copied for him, was not such as to raise doubts
about his orthodoxy. There is no evidence to imply that Wickins ever
suspected the transformation taking place under his eyes. Newton concealed his views so effectively that only in our day has full knowledge of
them become available.

❧

One of the points in his Arian credo, that only the Father has fore-
knowledge of future events, indicated another dimension of Newton's
early theological studies, the interpretation of the prophecies. Newton's
interest in the prophecies, Daniel and the Revelation of Saint John the
Divine, has been known since the publication of his *Observations upon the
Prophecies* shortly after his death. It has generally been assumed that the
work was a product of his old age, as the treatise published was. Nevertheless, references to the prophecies filled his early theological notebook.
Already in the 1670s, he believed that the essence of the Bible was the
prophecy of human history rather than the revelation of truths beyond
human reason unto life eternal. Already at that time he believed what he
asserted later about Revelation: "There [is] no book in all the scriptures
so much recommended & guarded by providence as this." He put that
belief into practice by composing his first interpretation of Revelation
while engaged in his earliest theological study. It proved to be more than a
passing interest. His first full discourse contains many insertions in later
hands, showing that he referred to it frequently. He composed numerous
revisions of it, one of which was probably the last thing on which he was at
work when he died more than fifty years later.

An introduction, which insisted on the critical importance of the
prophecies, opened the original treatise.

Having searched after knowledg in yᵉ prophetique scriptures [he began], I have
thought my self bound to communicate it for the benefit of others, remembering
yᵉ judgment of him who hid his talent in a napkin. For I am perswaded that this
will prove of great benefit to those who think it not enough for a sincere Christian
to sit down contented with yᵉ principles of yᵉ doctrin of Christ such as yᵉ Apostel

accounts the doctrin of Baptisms & of laying on of hands & of the resurrection of yᵉ dead & of eternall judgment, but leaving these & the like principles desire to go on unto perfection until they become of full age & by reason of use have their senses exercised to discern both good & evil. Hebr. 5.12

People should not be discouraged by past failures to understand these writings. God vouchsafed the prophecies for the edification of the church. They are not about the past. They were written for future ages. When the time comes, their meaning will be revealed in their fulfillment. Let people be warned by the example of the Jews, who have paid dearly for failing to recognize the promised Messiah. If God was angry with the Jews, He will be more angry with Christians who fail to recognize the Antichrist. "Thou seest therefore that this is no idle speculation, no matter of indifferency but a duty of the greatest moment." One cannot be too careful. The Antichrist comes to seduce Christians, and in a world of many religions, of which only one can be true "& perhaps none of those that thou art acquainted with," one must be circumspect in finding the truth. Already we perceive that Newton's interpretation of the prophecies was not unconnected with his Arianism.

In fact, Arianism – or perhaps its victorious opponent, trinitarianism – provided the key to Newton's interpretation. Interest in the prophecies had been rife in Puritan England, and a standard Protestant interpretation of Revelation, in which the Roman church inevitably played the role of the Beast, had emerged. Accepting the broad outlines of the Protestant interpretation, he altered its meaning to fit his new perception of Christianity. The Great Apostasy was no longer Romanism; it was trinitarianism.

Newton's interpretation differed from most contemporary ones in another way. Most students of the prophecies looked to them for understanding of contemporary events. Newton never did. Over the years he exhibited some, though not intense, interest in the date of the second coming, which he never placed in or near the late seventeenth century. Not once that I have seen did he attempt to relate the political history of England in his day to the prophecies. Rather, he riveted his attention on the fourth century, the crucial century of human history, when the Great Apostasy seduced humankind from worship of the one true God.

In Revelation, as Newton understood it, the opening of the first six seals, which represent successive periods of time, concern themselves with the history of the church until its definitive establishment within the empire in the reign of Theodosius. The seventh seal, within which the seven trumpets (which also represent successive periods of time) are included, began with the year 380. Until its conclusion at the sounding of the seventh trumpet, it portrays "one & the same continued Apostacy . . . . And indeed so notable are the times of this Apostacy y^t ye whole Apocalyps from ye fourth chapter seems to have been written for ye sake of it." Until then, trinitarian doctrines, though formulated by Athanasius, had been professed only by a few western bishops led by the pope. At that time, however, Theodosius became its patron and called the Council of Constantinople in 381 to ratify it.

The year 381 is therefore w^{th}out all controversy that in wch this strange religion of ye west wch has reigned ever since first overspread ye world, & so ye earth wth them that dwell therein began to worship ye Beast & his Image, y^t is ye church of ye western Empire & the afforesaid Constantinopolitan Counsel its representative . . . .

The mere thought of trinitarianism, the "fals infernal religion," was enough to fan Newton into a rage. With it had come the return of idolatry in a more degraded form, not the stately worship of dead kings and heroes in magnificent temples, but "ye sordid worship in sepulchres of ye Christian Divi . . . ye adoration of mean & despicable plebeians in their rotten reliques." Superstitions of every sort, fanned and spread by monks with feigned stories of false miracles, accompanied the new worship. "Idolaters," Newton thundered at them in the isolation of his chamber, "Blasphemers & spiritual fornicators . . . ." They pretended to be Christians, but the devil knew "that they were to be above all others ye most wicked wretched sort of people. . . . [The] worst sort of men that ever reigned upon the face of ye earth till that same time . . . ." The first six trumpets and the six vials of wrath corresponding to them represent successive invasions of the Empire – "like Furies sent in by the wrath of God to scourge ye Romans," repeated punishments of an apostate people who whored after false gods.

As the passion with which Newton expressed himself suggests, his early treatise on the prophecies was a very personal document. In his view, the triumph of trinitarianism had stretched beyond the limits of doctrine. It had won dominance by allying itself with base human motives, such as "covetousness & ambition . . . ."

It's plain therefore that not a few irregular persons, but ye whole clergy began at this time to be puft up, to set their hearts upon power & greatness more then upon piety & equity, to transgress their Pastoral office & exalt themselves above yͤ civil magistrate; not regarding how they came by praerogatives or of what ill nature or consequence they were, so they were but praerogatives, nor knowing any bounds to their ambition but impossibility & yͤ Imperial edicts.

Behind the specifically anti-Catholic element in Newton's denunciation, one can see a much broader condemnation. Restoration Cambridge supplied a more familiar example of covetousness and ambition in the church. Already Newton had separated himself in his daily life from the aspiring clerics about him. By embracing Arianism, he had expressed his scorn for their belief. Now, in his prophetic studies, he justified his rebellion by appealing to the divinely ordained course of human history. Men will call you a fanatic and heretic if you study the prophecies, he warned.

But yͤ world loves to be deceived, they will not understand, they never consider equally, but are wholly led by prejudice, interest, the prais of men, & authority of yͤ Church they live in: as is plain becaus all parties keep close to yͤ Religion they have been brought up in, & yet in all parties there are wise & learned as well as fools & ignorant. There are but few that seek to understand the religion they profess, & those that study for understanding therein, do it rather for worldly ends, or that they may defend it, then to examin whither it be true wͭʰ a resolution to chose & profess that religion wͨʰ in their judgment appears the truest. . . . Be not therefore scandalised at the reproaches of yͤ world but rather looke upon them as a mark of yͤ true church.

By the true church, for whom the prophecies are intended, he did not mean all who call themselves Christians "but a remnant, a few scattered persons which God hath chosen, such as without being led by interest, education, or humane authorities, can set themselves sincerely & earnestly to search after truth." There is no doubt that Newton placed himself among the select few. Some of his descriptions of the remnant have the poignancy of personal experience. Consider, he said in the earliest sketch of his interpretation, "yͤ Apostacies of yͤ Jewish Church under yͤ Law & particularly in Ahab's reign when yͤ undefiled part of yͤ Church so far disappeared yͭ yͤ Elijah thought himself yͤ only person left." Isolated in his chambers from the hedonism and triviality of Restoration Cambridge, Newton may have wondered if he was another Elijah, like the first almost the only true believer left.

A tension ran through the work. On the one hand, it contained a chiliastic flavor. Now at last the meaning of the prophecies was being revealed, and therefore the end had to be near, when the sealed saints who refused the mark of the Beast would resume their place at the head of the church. On the other hand, Newton was far from ready to identify anything about him as true apostolic Christianity. His deliberately chosen internal chronology set the day of the final trumpet two centuries away. About this, he was explicit. The beginning of the critical period of 1,260 years had come, not in 380 with the beginning of the seventh seal but at the close of the fourth trumpet, when the apostasy reached its peak, in the year 607. Whatever Newton's contempt for the society about him – perhaps because of his contempt – he did not think its conversion was near.

There is a further tension between his own passion and his dispassionate method. It is hardly surprising that a man whose first step in any new study was to organize his knowledge methodically should want to proceed similarly with Revelation. He complained of interpreters who, "wthout any such previous methodising of ye Apocalypse . . . twist ye parts of ye Prophesy out of their natural order according to their pleasure . . . ." In place of private fancy, Newton wanted certainty, for only then could the Bible be a clear rule of faith. His "Method" then would be the key to his approach. First, he would lay down rules of interpretation. Then he would give a key to the prophetic language, which would remove the liberty of twisting passages to private meanings. Third, he would compare the parts of Revelation with one another and digest them into order by the internal characters imprinted on them by the Holy Ghost, what he called opening scripture by scripture. One cannot miss the fact that this interpreter of the prophecies had trained his mind in the hard school of mathematics. It was not by accident that he proceeded in his exposition in a mathematical style, beginning with ten general "Propositions" (later renamed "Positions") followed by a larger number of particular ones, which drew upon them to build a demonstrative argument. Newton believed he had attained the certainty he sought. The original opening sentence of his treatise began: "Having searched & by the grace of God obteined knowledg . . . ."

※

There was a pressing problem, however, that could not wait until the late nineteenth century, the earliest point at which the chronology internal to

Newton's interpretation placed the second coming. In 1675, Newton would have to be ordained in the Anglican church or he would have to resign his fellowship. In the general laxity of Trinity, the requirement of ordination was one rule that was enforced. Four times within the previous decade, to fulfill university obligations, Newton had been willing to assert his orthodoxy under oath. To remain a fellow of the College of the Holy and Undivided Trinity, he would need to affirm his orthodoxy one last time, in ordination. By 1675, however, the holy and undivided trinity itself stood in the way. "If any shall worship the Beast and his image and receive his mark in his forehead or in his hand," Revelation told him, "the same shall drink of the wine of the wrath of God . . . ." Newton did not doubt the literal truth of the Word. Accept ordination he could not.

More than the fellowship was at stake. If Newton despised the society of Trinity, the material support that the college provided, in a location which assured his access to the world of learning, was the bedrock of his existence. Perhaps he could have held his chair without the fellowship and remained in Cambridge, although I do not know of another such case. The problem was secrecy. Questions were bound to be asked. In itself, ordination did not entail duties. It did not entail ecclesiastical appointment. Why would anyone in Newton's position, someone who intended to stay on in Cambridge, celibate, surrender a fellowship worth £60 a year for no reason at all? Or rather, what was one to conclude about the true reason that led a man in such a position to refuse ordination? Questions were bound to be asked. Questions were exactly what Newton had to avoid. Heresy was grounds for ejection from his chair, as William Whiston later learned. Newton's particular heresy was grounds for ostracism from polite society, as Whiston also learned. It is impossible even to imagine what the consequences for Newton would have been had he been branded a moral leper in 1675. As the deadline approached, his career faced another crisis.

In fact, there was a possible means of escape. Any statute could be set aside by a royal dispensation. Late in 1674, Francis Aston, the fellow to whom Newton wrote in 1668, attempted to obtain a dispensation to free himself from the obligation of ordination. A letter from Barrow, who was then master of Trinity, to the Secretary of State, Joseph Williamson, on 3 December 1674, presented the college's case against a dispensation. It would destroy succession and subvert the principal end of the college, which was the breeding of clerics. He was sure the senior fellows would refuse to accept it. Aston did not obtain a dispensation. The draft of a letter from Newton to "Sr Alexander," which indicates that Newton had been involved in Aston's attempt, survives. Sir Alexander was probably Sir Alexander Frazier, physician and confidant of Charles II, whose son, Charles Frazier, was elected to a fellowship in Trinity in 1673. In the letter, Newton thanked Sir Alexander for including him in the proposal for a dispensation, which, he said, had been successfully opposed by the college. The vice-master objected strongly, and the seniors followed him. They said that granting a dispensation would hinder succession (almost exactly Barrow's phrase), and furthermore they did not wish to recede so far from fundamental statutes of the college.

By early 1675, Newton had given up hope. He wrote to Oldenburg in January requesting that the Royal Society excuse him from payments, as Oldenburg had offered two years before. "For ye time draws near y^t I am to part wth my Fellowship, & as my incomes contract, I find it will be convenient that I contract my expenses."

At the last moment, the clouds lifted. Less than a month after his letter to Oldenburg, Newton went to London. By 2 March, Secretary Coventry, with the rationale that His Majesty was willing "to give all just encouragement to learned men who are & shall be elected to ye said Professorship," sent the draft of a dispensation to the Attorney General for his opinion. On 27 April, the dispensation became official. By its terms, the Lucasian professor was exempted from taking holy orders unless "he himself desires to . . . ." We know nothing of the factors behind the recorded events. The presence of Humphrey Babington among the senior fellows could not have injured Newton's prospects. Nevertheless, it appears more likely that on this occasion it was Isaac Barrow who rescued Newton from threatened oblivion. A dispensation was a royal act, and Barrow was the one who had the ear of the court. Although Newton's letter to Sir Alexander stated that he was involved in a joint endeavor to gain a dispensation, Barrow's letter to Secretary Williamson in December concerned Aston alone. In the letter to Sir Alexander about the proposal, Newton specifically stated that the master "received it kindly . . . ." We can only speculate on what passed between Barrow and Newton. Barrow was deeply committed to the church, and it is hard to believe that he would have acquiesced in a plea of Arianism. It is not hard to believe that he would have acquiesced in an argument that Newton had no vocation for the ministry. Barrow understood Newton's worth, and he valued learning. Moreover, he would have recognized that Newton, in contrast to Aston, would not set a precedent. As Lucasian professor, he was unique. The dispensation was granted to the Lucasian professorship in perpetuity, not to Isaac Newton, fellow of Trinity. It was probably Barrow's last service to his protégé.

Once more a crisis that threatened Newton's scientific career before it had reached fruition passed. Now at last, despite gross heresy sufficient to make him a pariah, he had surmounted the final obstacle and found himself secure in his sanctuary. And he had demonstrated a new facet of his genius: He could have his cake and eat it too.

7

Years of Silence

B Y THE END OF 1676, as absorbed in theology and alchemy as he was distracted by correspondence and criticism on optics and mathematics, Newton had virtually cut himself off from the scientific community. Oldenburg died in September 1677, not having heard from Newton for more than half a year. Newton terminated his exchange with Collins by the blunt expedient of not writing. It took him another year to conclude the correspondence on optics, but by the middle of 1678, he succeeded. As nearly as he could, he had reversed the policy of public communication that he began with his letter to Collins in 1670 and retreated to the quiet of his academic sanctuary. He did not emerge for nearly a decade.

Humphrey Newton sketched a few facets of Newton's life as he found it in the 1680s. Newton enjoyed taking a turn in his garden, about which he was "very Curious . . . not enduring to see a weed in it . . . ." His curiosity did not rise to the level of dirtying his hands, however; he hired a gardener to do the work. He was careless with money; he kept a box filled with guineas, as many as a thousand, Humphrey thought, by the window. Humphrey was not sure if it was carelessness or a deliberate ploy to test the honesty of others – primarily Humphrey. In the winter, he loved apples, and sometimes he would have a small roasted quince. Not much in the account suggested leisure, however. The Newton Humphrey found had immersed himself in unremitting study to the extent that he grudged even the time to eat and sleep. During five years, Humphrey saw him laugh only once, and John North, master of Trinity from 1677 to 1683, feared that Newton would kill himself with study.

These were disastrous years for the college. In 1675, when it was still too soon to recognize the decline in numbers that had just set in, Isaac

Barrow committed the college to the construction of an extravagant library. The magnificent Wren structure became an ornament to the entire university, as Barrow intended, but its burden of expense hamstrung the college for two decades until its completion in 1696. Two disastrous masters who succeeded Barrow left the college adrift at precisely the time when it most needed vigorous leadership. The extraordinary longevity of the seniors during this time meant general senility among those who might have supplied the masters' wants. By the late 1680s, the financial crisis of the library had spread into financial chaos for the college as a whole, and dividends began to fail.

The problems of the college had to touch Newton. He did not withdraw completely from academic life. He voted in the senate in various university elections, for example. We know of the contribution he made for the library, however, and of his loan to help finance it only from the college records. He lived in Trinity. He never gave the college his heart.

Late in the spring of 1679, Newton's mother died. As Conduitt heard the story, her son Benjamin Smith was seized with a malignant fever in Stamford. She went to nurse him and caught the fever. Newton in turn went to nurse his mother,

sate up whole nights with her, gave her all her Physick himself, dressed all her blisters with his own hands & made use of that manual dexterity for wch he was so remarkable to lessen the pain wch always attends the dressing the torturing remedy usually applied in that distemper with as much readiness as he ever had employed it in the most delightfull experiments.

Despite his ministrations, she died. An entry in the parish record of Colsterworth recorded her death. "Mrs Hannah Smith, wid. was burried in woollen June ye 4th 1679." Filial love is an attractive quality for a variety of reasons, some moral, some psychoanalytical. I trust it is not too cynical to note that during the twelve years following his return to Cambridge after the plague, Newton is known to have paid three visits to Woolsthorpe. On three other absences from Cambridge, he probably returned home also. More vigorous displays of filial affection have been recorded.

Death imposes practical demands, and his mother's death led to Newton's longest stay in Woolsthorpe, the plague years aside, after he was called home from grammar school twenty years before. He was both heir and executor, and it took him virtually all of the rest of 1679 to put his affairs in order. He devoted nearly as much time to his estate in 1679 as he had to his mother in all his visits in the previous twelve years combined. According to the Steward's Book of Trinity College, in addition to the summer, he was absent for nearly four months of the year that began on Michaelmas 1679. Much of the time was spent in Woolsthorpe. Part of the problem was a tenant, probably Edward Storer, one of the stepsons of the apothecary Clark, with whom Newton had lived in Grantham. At least, Edward Storer and his sons were Newton's tenants eight years later, and their unsatisfactory relations then had accumulated during a tenure of unspecified length.

Not long thereafter, Newton sustained another loss. In 1683, after an extended period of nonresidence during which he came to the college only one or two weeks a year, Wickins decided to resign his fellowship. He visited the college for three weeks in March 1683; though he remained a fellow beyond Michaelmas 1684, he never returned to Trinity again. Probably Wickins had already been inducted into the rectory of Stoke Edith, Hereford. The Foley family, into which a sister of the master, John North, married, controlled the living; there is every reason to think that North recommended one of his fellows, who was clearly on the prowl, to them. In this at least, North built well. A Wickins occupied the rectory of Stoke Edith under the patronage of a Foley for more than a century. Undoubtedly, it was the decision to marry and beget that progeny which led Wickins to resign his fellowship.

Newton's relation to Wickins remains a mystery. Despite a friendship of twenty years, we know nothing about it aside from the anecdote of their meeting as undergraduates and Wickins's service as amanuensis, to which numerous papers in his hand testify. The blank that followed Wickins's departure wraps the mystery in an enigma. After Newton's death, Robert Smith of Trinity wrote to Wickins's son Nicholas for information about Newton. Nicholas Wickins replied that his father, who had died earlier, had set out once to collect everything related to Newton that he possessed. He had transcribed into a notebook three short letters, so uninformative that Nicholas did not bother to send them, and he had in addition four or five very short letters in which Newton had merely forwarded dividends and rent. He said nothing about dates, but most of the second group at least had to have come before 1683. He went on to tell the story of how his father met Newton, and he added three other superficial anecdotes he had heard. He concluded with the information that through his father and through himself Newton had supplied Bibles to the poor in the parish of Stoke Edith, apparently their single topic of communication after 1683. There is an unsatisfactory tone about the letter, as though someone, Wickins or his son, were concealing something. I am unable to imagine a close friendship of twenty years leaving so little residue unless it ended in a breach. Newton's draft of the single letter to Wickins that we have, brief to the point of being curt, implied some barrier which he could not transcend. Written sometime between 1713 and 1719, it responded to a request for more Bibles to distribute in Stoke Edith; Newton indicated that he would send them via Wickins's patron, Thomas Foley. He brushed off Wickins's attempt at a friendly exchange. "I am glad to heare of your good health, & wish it may long continue, I remain . . . ." On the same day, 28 March 1683, that Wickins left Trinity for the last time, Newton also left. He returned on 3 May and left again for another week on 21 May. We have no idea where he went.

Later that year, he arranged to have a young man from his own grammar school in Grantham, Humphrey Newton, come to live with him as his amanuensis. According to Stukeley, who conversed with Humphrey in 1727 and 1728, he was "under Sr Isaac's tuition . . . ." This suggests the status of sizar, though Humphrey Newton was never admitted to Trinity. In the following five years, as he lived in Newton's chamber like Wickins before him, he copied out extensive writings, at first primarily on theology and mathematics. About a year after he arrived, a visit from Edmond Halley set off a new investigation which ensured Humphrey's immortality; it was Humphrey who transcribed the copy from which the *Principia* was printed. Years later, when he married as an old man, he named his son Isaac after his "dear deceased Friend . . . ."

ॐ

Theological study occupied much of Newton's time during the years of silence. In the late 1670s, he began a history of the church, concentrating on the fourth and fifth centuries, which repeated the themes of his interpretation of Revelation. Athanasius played the role of the villain, of course. Some passages functioned as first drafts of his treatise from the same period, "Paradoxical Questions concerning the morals & actions of Athanasius & his followers," in which Newton virtually stood Athanasius in the dock and prosecuted him for a litany of sins too long to enumerate here. In these papers, the passion evident in his earlier interpretation of Revelation rose to a new level of intensity as Newton sought to show, not only that Athanasius was the author of "the whole fornication" – that is, of trinitarianism, "the cult of three equal Gods" – but also that Athanasius was a depraved man ready even to use murder to promote his ends.

Although much has been written about the influence of the Cambridge Platonist Henry More on Newton, and much speculated on the basis of their common origin in Grantham, concerning their mutual contact in Cambridge, we know very little that is concrete about their relations. We do know, however, that Newton cited More's contribution to understanding Revelation from the beginning of his own study, and we do know that the two men discussed the prophecy. It was in connection with such a discussion that More drew one of the most revealing sketches of Newton. Many persons repeated anecdotes about Newton's neglecting his meals. Only More fully caught him in a state of ecstasy. Writing to John Sharp in 1680, he mentioned how well he and Newton agreed about the Revelation.

For after his reading of the Exposition of the Apocalypse which I gave him, he came to my chamber, where he seem'd to me not onely to approve my Exposition as coherent and perspicuous throughout from the beginning to the end, but (by the manner of his countenance which is ordinarily melancholy and thoughtfull, but then mighty lightsome and chearfull, and by the free profession of what satisfaction he took therein) to be in a maner transported.

More went on to describe their areas of disagreement, and he did not hesitate to call Newton's identification of the seven vials and the seven trumpets "very extravagant." He assumed that Newton would see his error as soon as he read More's exposition more carefully. In this, he rather underestimated Newton's tenacity. There is nothing in the letter to imply that Newton had even hinted at his private understanding of the great apostasy. The two men must have maintained some intimacy. When More died six years later, Newton was one of only five outside the fellowship of Christ's College to whom he bequeathed a funeral ring.

While Humphrey was with him, Newton also undertook a new theological venture which became henceforth the vehicle for his heterodox theological opinions. In its implications "Theologiae gentilis origines philosophicae" (The Philosophical Origins of Gentile Theology) was more radical than any Arian statement he composed during the 1670s. The "Origines" started with the argument that all the ancient peoples worshiped the same twelve gods under different names. The gods were divinized ancestors – in fact Noah, his sons, and his grandchildren – though as this religion passed from people to people, each used it to its own ends by identifying the gods with its own early kings and heroes. Nevertheless, common characteristics distinguished the corresponding gods of all the ancient peoples.

The number twelve derived from the seven planets, the four elements, and the quintessence. Newton argued that peoples had identified their most eminent ancestors with the most eminent objects that nature presents. Even in the seventeenth century, Galileo named the newly discovered satellites of Jupiter after his benefactors; so in the earliest ages of human society, humans had identified their most prominent ancestors with the heavenly bodies, assumed that their souls transmigrated to those bodies, and begun to ascribe divine powers to the souls. Hence, as the title of the treatise proclaimed, gentile theology had its origin in natural philosophy. Chapter 1 asserted the same thing: "Gentile theology was philosophical and dependent on the astronomy and physical science of the system of the world." He frequently called it "astronomical theology" and similar names.

Egypt was the original home of gentile theology. Here Noah settled after the flood, and from Egypt other lands were settled when the sons of Noah fought over the inheritance and separated. The Egyptians first developed sidereal theology, which looked back to their own ancestors. They taught it to other peoples, who had the same ancestors but remodeled the gods in order to enhance their own self-esteem.

Newton was convinced that gentile theology represented a falling away from true religion.

It cannot be believed, however, that religion began with the doctrine of the transmigration of souls and the worship of stars and elements: for there was another religion more ancient than all of these, a religion in which a fire for offering sacrifices burned perpetually in the middle of a sacred place. For the Vestal cult was the most ancient of all.

He cited evidence to prove that a similar worship had been the most ancient in Italy, Greece, Persia, and Egypt, among other places. When

Moses instituted a perpetual flame in the tabernacle, he restored the original worship "purged of the superstitions introduced by the Egyptians." Such had been the worship of Noah and his sons. In his turn, Noah had learned it from his ancestors. It was indeed the true worship instituted by God. "Now the rationale of this institution was that the God of Nature should be worshiped in a temple which imitates nature, in a temple which is, as it were, a reflection of God. Everyone agrees that a Sanctum with a fire in the middle was an emblem of the system of the world." Human beings are "ever inclined to superstitions," however. The Egyptians corrupted the true worship of God by creating false gods from their ancestors; other peoples were only too ready to learn degenerate practices from them.

Another name for the false worship was idolatry. To Newton, idolatry represented the fundamental sin. Because true worship took place in a sanctum that represented God's creation, it followed directly that its corruption involved a corruption of natural philosophy as well; so Newton believed. The original temple, with the fire in the center, illuminated by seven lamps representing the planets, symbolized the world.

The whole heavens they recconed to be ye true & real temple of God & therefore that a Prytanaeum [Sanctum] might deserve ye name of his Temple they framed it so as in the fittest manner to represent the whole systeme of the heavens. A point of religions then wch nothing can be more rational. . . . So then twas one designe of ye first institution of ye true religion to propose to mankind by ye frame of ye ancient Temples, the study of the frame of the world as the true Temple of ye great God they worshipped. . . . So then the first religion was the most rational of all others till the nations corrupted it. For there is no way (w^thout revelation) to come to ye knowledge of a Deity but by ye frame of nature.

Geocentric astronomy accompanied the spread of false religion. It was not by accident that Ptolemy was also an Egyptian.

So far, it was possible to fit the "Origines" into an orthodox Christian setting. Nevertheless, the orthodox would have found other elements of the treatise inadmissible. Though he admitted the Mosaic account, he checked it against other ancient testimony, such as the Phoenician chronicler Sanchuniathon, and the Chaldean Berossus, in effect treating it on a par with pagan testimony. More significant was the implicit deemphasis of the role of Christ, a step which came readily enough to an Arian. Instead of the agent of a new dispensation, Christ was a prophet, like Moses before him, sent to recall humankind to the original true worship of God. As he revised the "Origines," Newton set down a number of chapter headings, the last of which was for chapter 11. "What was the true religion of the sons of Noah before it began to be corrupted by the worship of false Gods. And that the Christian religion was not more true and did not become less corrupt." In this setting, trinitarianism with its encouragement of the worship of saints and martyrs, indeed with its worship of Christ as God, took on a new meaning. What was trinitarianism but the latest manifestation of the universal tendency of humankind to superstition and idolatry? Through Athanasius, Egypt once again played its nefarious role as the corrupter of true religion. By universalizing the Christian experience of the first four centuries, Newton denied it any unique role in human history. The Christian religion rightly understood was not more true than the religion of the sons of Noah, which was founded upon the recognition of God in His creation.

Perhaps it is not altogether surprising that Humphrey Newton remembered a man who, despite what we know of his intense theological concern, did not trouble himself much with the established forms of public worship. Though he usually went to St. Mary's on Sundays, he seldom went to chapel. Morning chapel fell during the time he slept, and vespers during the time he studied, "so y^t He scarcely knew ye Hour of Prayer." It is perhaps also not surprising that he kept the "Origines" to himself, so that its very existence has become known only in this day, nearly three hundred years after he conceived it.

&.

Theology was not Newton's only occupation during these years. The manuscript remains testify that alchemy vied with theology to command his attention. He continued to receive unpublished alchemical manuscripts to copy, and the letter from Francis Meheux in 1683 provides one of our most explicit pieces of evidence of his contact with alchemical circles. Newton did more than read alchemy. He also experimented extensively. He left behind dated experimental notes beginning in 1678 and extending, with a break in the late 1680s, nearly to the time of his departure from Cambridge. His experimentation left a deep impression on Humphrey Newton. We know from surviving manuscripts that Humphrey copied extensive mathematical and theological manuscripts for Newton.

He also transcribed the entire *Principia,* the composition of which occupied at least half of his stay in Cambridge. Neither of the first two activities figured in Humphrey's recollections at all, and he gave only two sentences to the *Principia.* Chemical activities bulked very large in his memory, however. He recalled that Newton slept very little,

especially at Spring & Fall of yᵉ Leaf, at wᶜʰ Times he used to imploy about 6 weeks in his Elaboratory, the Fire scarcely going out either Night or Day, he siting up one Night, as I did another, till he had finished his Chymical Experiments, in yᵉ Performances of wᶜʰ he was yᵉ most accurate, strict, exact: What his Aim might be, I was not able to penetrate into, but his Pains, his Diligence at those sett Times, made me think, he aim'd at something beyond yᵉ Reach of humane Art & Industry.

He employed himself "with a great deal of satisfaction & Delight" in his laboratory, Humphrey added. He built and remodeled his own furnaces. The laboratory was very "well furnished with Chymical Materials, as Bodyes, Receivers, Heads, Crucibles &c, wᶜʰ was made very little use of, yᵉ Crucibles excepted, in wᶜʰ he fus'd his Metals." Sometimes, though very seldom, he would look in an "old mouldy Book" which Humphrey thought was Agricola, *De metallis* [*De re metallica*], "The transmuting of metals, being his Chief Design, for wᶜʰ Purpose Antimony was a great Ingredient." Although the sparsity of references to Agricola among Newton's papers suggests that Humphrey supplied that detail by grasping at a well-known work that he thought was appropriate, the extensive experimental notes that Newton left behind from this period do support the principal elements of his account.

In the spring of 1681, Newton's experimentation reached a climax. In the midst of his experimental notes, which he set down in English, he entered in Latin two paragraphs that were obviously not experimental notes but appear to have been meant as interpretations of them. They employ the same mythological imagery that appears in the alchemical manuscripts. An exultant air pervades the paragraphs; one can almost hear the triumphant "Eureka!" ring through the garden.

May 10 1681 I understood that the morning star is Venus and that she is the daughter of Saturn and one of the doves. May 14 I understood ⫤ [the trident?]. May 15 I understood "There are indeed certain sublimations of mercury" &c as also another dove: that is a sublimate which is wholly feculent rises from its bodies white, leaves a black feces in the bottom which is washed by solution, and mercury is sublimed again from the cleansed bodies until no more feces remains in the bottom. Is not this very pure sublimate —✳? [sophic sal ammoniac?]

The following experimental note, which intervened before the second interpretive paragraph, employed "sophic ✳, [sal ammoniac]" which I take as support for my interpretation of the symbol —✳.

May 18 I perfected the ideal solution. That is, two equal salts carry up Saturn. Then he carries up the stone and joined with malleable Jupiter [as much as one wants to say "tin" here, Newton did write *Jove* instead of inserting the symbol ♃] also makes ✳ [sophic sal ammoniac?] and that in such proportion that Jupiter grasps the scepter. Then the eagle carries Jupiter up. Hence Saturn can be combined without salts in the desired proportions so that fire does not predominate. At last mercury sublimate and sophic sal ammoniac shatter the helmet and the menstruum carries everything up.

An entry in Newton's other set of laboratory notes, also in Latin, with the date "July 10" but no year, appears to belong to the same climactic experiments of 1681. "I saw sophic sal ammoniac. It is not precipitated by salt of tartar." From this time, Newton used mostly sophic sal ammoniac, which he called a number of names such as "✳ prep" (prepared sal ammoniac) and sometimes distinguished from vulgar sal ammoniac, in his experimentation. In 1682, he contrasted "✳ prepared with ♁ [antimony]" to "✳ without ♁," and once he referred to "the green lion (or our ✳)."

At some point, he crossed out the two triumphant paragraphs from May, probably indicating that the triumph had dissolved into failure. Nevertheless, the disillusionment was apparently not complete, for he continued from this time to employ sophic sal ammoniac in his experimentation. And ten years later, when he composed his most important alchemical treatise, he employed all the imagery that the paragraphs embodied.

In 1684, Newton squeezed in another exultant note between two lines of his experimental notes. "Friday May 23 I made Jupiter fly on his eagle." By May 1684, Humphrey Newton was helping to tend the furnaces. "Nothing extraordinary, as I can Remember, happen'd in making his experiments," he recounted to Conduitt, "wch if there did, He was of so sedate & even Temper, y^t I could not in ye least discern it." Newton had his own way of expressing his elation. He shouted his eurekas to himself. Nevertheless, one might think that Jupiter flying on his eagle near the east end of Trinity chapel would have attracted Humphrey's attention. Or he might have noticed that Newton got excited enough with his experiments to forget, as his notes reveal, what day of the week it was.

During the later 1670s, following his response to Robert Boyle's letter about a special mercury in the *Philosophical Transactions*, Newton entered into direct correspondence with Boyle. Late in 1676, Boyle sent two copies of his recent book via Oldenburg, one for Newton and one he was asked to deliver to Henry More. Newton immediately sent his thanks through Oldenburg because he did not have Boyle's address. Three months later, again in contrast to his usual reticence, he volunteered a comment on another article by Boyle in the *Philosophical Transactions*. Some time before 28 February 1679, his manifest effort to establish a correspondence bore fruit. Although the letter of that date is the first that survives, it referred to earlier communication. Newton's letter has been known since it was printed in the eighteenth century, and it has often been republished as a typical example of the mechanical philosophy at work. In fact, it shows that Newton had moved still further from the stance of orthodox mechanical philosophy, and its content suggests that alchemy had played the crucial role in moving him.

The book that Boyle presented to Newton late in 1676 was probably *Experiments, Notes, &c. about the Mechanical Origine or Production of Divers Particular Qualities*. Newton's letter of 1679 commented implicitly on some of the central arguments of the book, arguments about the dissolution, precipitation, and volatilization of substances.

Honoured S^r [he began]

I have so long deferred to send you my thoughts about ye Physicall qualities we spake of, that did I not esteem my self obliged by promise I think I should be ashamed to send them at all. The truth is my notions about things of this kind are so indigested y^t I am not well satisfied my self in them, & what I am not satisfied in I can scarce esteem fit to be communicated to others, especially in natural Philosophy where there is no end of fansying.

He concluded the letter by saying that Boyle could "easily discern whether in these conjectures there be any degree of probability, wch is all I aim at. For my own part I have so little fansy to things of this nature that had not your encouragement moved me to it, I should never I think have thus far set pen to paper about them." We can judge how seriously to take such deprecations by the fact that he composed four thousand words on the fancies he fancied so little.

In his book, Boyle had cited many chemical reactions in which heat appeared. Chemists referred the heat, he said, to violent antipathies between substances; as a mechanical philosopher, he ascribed it to the motion among the particles of the substances. By 1679, as his experimental notes testify, Newton had had first-hand experience with many such reactions. Indirectly, he raised with Boyle the question of the cause of motion among the particles of previously cold substances. He himself explained the motion by an "endeavour . . . bodies have to recede from one another . . . ." When a particle is separated from a body in dissolution, it is accelerated by the endeavor to recede "so y^t ye particle shall as [it] were wth violence leap from ye body, & putting ye liquor into a brisk agitation, beget & promote y^t heat we often find to be caused in solutions of Metals." He traced the phenomena of surface tension and expansion of air, which had fascinated him since his introduction to natural philosophy, to the same endeavor to recede. Under certain conditions, bodies also endeavor to approach one another, and he argued that this endeavor causes the cohesion of bodies. He derived both endeavors, in turn, from mechanisms within the aether, with the positing of which the letter began. Nevertheless the very language of "endeavors" was a step toward a radically different explanation.

The letter concerned itself primarily with the causes of solubility and volatility, two of the qualities to which Boyle had directed his treatise. Chemists, Boyle had argued, generally attributed solubility to a certain sympathy between the substance in question and its menstruum. As a mechanical philosopher, he was unable to understand what such a sympathy could be unless it were solely a matter of the sizes and shapes of particles and pores. Newton did not use the word *sympathy*, but he did assert that "there is a certain secret principle in nature by wch liquors are sociable to some things & unsociable to others." He expressly denied that it has anything to do with the sizes (and by implication the shapes) of pores and particles. He had mentioned such a principle in his "Hypothesis of Light." Its existence and its role in nature provided the central argument of the letter to Boyle. By inference, an unsociability between aether and gross bodies caused the rarity of aether in the pores of bodies, the basis of his explanation of the endeavors to approach and recede and of the new explanation of gravity that he offered at the end of the letter. The basis of the secret principle of sociability, in turn, was entirely chemical. That is, he justified his assertion of it by chemical phenomena. Water does not mix with oil but does mix readily with spirit of wine and with salts. Water sinks into wood, but quicksilver does not. Quicksilver sinks into metals, but water does not. Aqua fortis dissolves silver but not gold; aqua regis dissolves gold but not silver. So also he illustrated by chemical phenomena the principle of mediation by which unsociable substances are brought to mix. Molten lead does not mix with copper or with regulus of Mars, but with the mediation of tin it mixes with either. With the mediation of saline spirits, water mixes with metals; that is, acids (water impregnated with saline spirits) dissolve metals. He employed the same arguments, the endeavor to recede and the principle of sociability, to explain volatization; and in passing, he dropped comments about metallic exhalations from the earth and metallic particles as the constituents of true permanent air. The letter to Boyle opened with a set of five suppositions about the aether that gave it the appearance of standard mechanical philosophy. The centrality in it of the principle of sociability, however, transformed it into an assertion of the insufficiency of the mechanical philosophy of nature for the explanation of chemical phenomena.

Meanwhile, in close connection with the letter to Boyle in 1679 Newton began a treatise known, from the titles of its two chapters, as "De aere et aethere" ("Concerning the Air and the Aether"). From its content, it appears to have been an effort to expound the same phenomena in the form of a systematic treatise. Whereas the letter to Boyle began with the postulation of an aether, "De aere" began with observed phenomena of the air, primarily its capacity to expand, one of the critical phenomena that had seized Newton's attention in 1664–5. The absence of external pressure allows air to expand. Heat causes it to expand. The presence of other bodies also causes it to expand. To justify the last assertion, Newton pointed to capillary phenomena, which arise from differences in pressure "because the air seeks to avoid the pores or intervals between the parts of these bodies . . . ." Indeed, he found that in general bodies seek to avoid one another, and he cited in justification, among other things, the phenomenon of surface tension. Various explanations of these phenomena might be offered, he continued.

But as it is equally true that air avoids bodies, and bodies repel each other mutually, I seem to gather rightly from this that air is composed of the particles of bodies torn away from contact, and repelling each other with a certain large force.

Only in such terms, he argued, can we understand the immense expansions that air can undergo until it fills volumes a thousand times its normal ones,

which would hardly seem to be possible if the particles of air were in mutual contact; but if by some principle acting at a distance [the particles] tend to recede mutually from each other, reason persuades us that when the distance between their centers is doubled the force of recession will be halved, when trebled the force is reduced to a third and so on.

As in the letter to Boyle, he went on to discuss the generation of air by various processes in nature.

At one point, Newton started to speculate on the possible causes of the repulsion between bodies. He crossed the paragraph out, presumably because chapter 1 expounded phenomena only and he intended to reserve such discussion for another place. The third possible explanation he offered before he canceled the passage had a familiar look. "Or it may be in the nature of bodies not only to have a hard and impenetrable nucleus but also a certain surrounding sphere of most fluid and tenuous matter which admits other bodies into it with difficulty." That is, the alchemical hermaphrodite, sulphur surrounded by its mercury, offered a model to explain the universal property that all bodies possess to act upon one another at a distance.

Chapter 2 of the treatise bore the title "De aethere," and in it Newton started to discuss the generation of aether by further fragmentation of aerial particles into smaller pieces. Apparently he intended to use the aether, as he had in the letter to Boyle, to explain what he had there called the endeavor to recede. First, he began to list the evidence for the existence of an aether. He cited electric and magnetic effluvia and the saline spirit that passes through glass and causes metals calcined in sealed vessels to increase in weight. He also cited the fact that a pendulum in an evacuated receiver stops almost as soon as one in the open air. Already, when he composed the "Quaestiones," this experiment had seemed to Newton to demonstrate the presence of a resisting medium in the receiver, and he had cited it as evidence for the aether in the "Hypothesis of Light." Newton did not proceed very far with the chapter on the aether. After a few lines, in the middle of a sentence in the middle of a page, he stopped, and he never took it up again. There are many possible reasons why Newton broke off the treatise. Mr. Laughton may have called; he may have gone to dinner and forgotten the paper when he returned; or he may have paused to reflect on the argument itself and realized that he was committing himself to an infinite regression in which a further aether would be required to explain the aether that explained the properties of air, a third aether to explain the second, and so on.

He may also have begun to think more deeply about the evidence of the pendulum experiment. Newton had believed that aether resists the motion of bodies differently from air. Air encounters only the surface of a body, whereas aether penetrates its pores and strikes against all of its internal surfaces as well. Newton had accepted the experiment of the pendulum in a void from Boyle's *Spring of the Air*, one of the first books on the new natural philosophy that he read. Could the issue be more complicated than he had thought? Could he refine the pendulum experiment to make it more demonstrative? In the *Principia*, Newton described such a refined experiment, which he had to relate from memory because he had lost the paper on which he recorded it. He did not date the experiment, but it had to have been after 1675, when he cited the earlier experiment in the "Hypothesis," and if my dating of "De aere et aethere" is correct, at least as late as 1679. If he performed it as late as 1685 in connection with the *Principia*'s composition, it is difficult to imagine that he would have lost the paper. It must have impressed him profoundly, because he remembered its details clearly. Newton constructed a pendulum eleven feet long. To minimize extraneous resistance, he suspended it from a ring that hung on a hook filed to a sharp edge. The first time he tried it, the hook was not strong enough, and it introduced resistance by bending to and fro. Newton was careful enough to notice it – and of course to correct it with a stronger hook. For the bob of the pendulum, he used a hollow wooden box. He pulled it aside six feet and carefully marked the places to which it returned on the first, second, and third swings. To be certain, he did it several times. Then he filled the box with metal, and by careful weighing – in which he included the string around the box, half the length of the string, and even the calculated weight of the air inside the box – he determined that the filled box was seventy-eight times heavier than the empty one. The increased weight stretched the string, of course; he adjusted it to make the length equal to the original. He pulled it aside to the same starting point and counted how many swings it took to damp down to the marks for the empty one. The pendulum required seventy-seven swings to reach each successive mark. Because the box filled had seventy-eight times as much inertia, the resistance full apparently bore the ratio of 78/77 to the resistance empty. By calculation, he concluded that this ratio corresponded to a resistance on the internal surfaces that was 1/5,000 the resistance on the external surface.

This reasoning [he concluded] depends upon the supposition that the greater resistance of the full box arises from the action of some subtle fluid upon the included metal. But I believe the cause is quite another. For the periods of the oscillations of the full box are less than the periods of the oscillations of the empty box, and therefore the resistance on the external surface of the full box is greater than that of the empty box in proportion to its velocity and the length of the spaces described in oscillating. Hence, since it is so, the resistance on the internal parts of the box will be either nil or wholly insensible.

In the letter to Boyle, Newton had argued that mechanical principles are inadequate to explain all phenomena. Now he had demonstrated to his own satisfaction that the aether, the deus ex machina that made mechanical philosophies run, does not exist. Driven, as it seems to me, primarily by the phenomena that he observed in alchemical experimentation and encouraged by concepts that he encountered in alchemical study, Newton appeared to be poised in 1679 on the brink of a further break with the mechanical philosophy, which would have major impact on his future career.

≈

Other questions in natural philosophy retained their power to excite Newton. He did not turn to them spontaneously during the years of silence, but others obtruded them upon his consciousness. Despite his withdrawal, a variety of men turned to him with questions, and the questions usually stimulated more than a simple reply. Newton's letter to Boyle early in 1679 was an answer to one such question; "De aere et aethere" was his more extended private response. Late in 1679, immediately upon his return from Woolsthorpe, another intrusion burst in upon him, this one a letter from Robert Hooke. Writing as Oldenburg's successor as secretary of the Royal Society, Hooke invited Newton to resume his earlier correspondence. He passed on a few items of information, and he specifically asked for Newton's opinion of his own hypothesis that planetary motions are compounded of a tangential motion and "an attractive motion towards the centrall body . . . ."

Hooke was referring to a remarkable paragraph that had concluded his *Attempt to Prove the Motion of the Earth* (1674; republished in 1679 in his *Lectiones Cutlerianae*). There he had mentioned a system of the world he intended to describe, a system embodying a concept not far removed from universal gravitation. Perhaps more important, he correctly defined, for the first time anyone had done so, the dynamic elements of orbital motion. Hooke said nothing about centrifugal force. Orbital motion results from the continual deflection of a body from its tangential path by a force toward some center. Newton's papers reveal no similar understanding of circular motion before this letter. Every time he had considered it, he had spoken of a tendency to <u>recede</u> from the center, what Huygens called centrifugal force; and like others who spoke in such terms, he had looked upon circular motion as a state of equilibrium between two equal and opposing forces, one away from the center and one toward it. Hooke's statement treated circular motion as a disequilibrium in which an unbalanced force deflects a body that would otherwise continue in a straight line. It was not an inconsiderable lesson for Newton to learn.

In his reply, written the day after his return to Cambridge, Newton began by declining the proffered correspondence. For the past six months family affairs in Lincolnshire had occupied him so fully that he had had no time for philosophical speculation.

And before that, I had for some years past been endeavouring to bend my self from Philosophy to other studies in so much y^t I have long grutched the time spent in y^t study unless it be perhaps at idle hours sometimes for a diversion. . . . And having thus shook hands wth Philosophy, & being also at present taken of wth other business, I hope it will not be interpreted out of any unkindness to you or ye R. Society that I am backward in engaging my self in these matters . . . .

But he was not quite able to leave it at that, and he allowed himself to suggest an experiment to reveal the diurnal rotation of the earth. The classic objection against diurnal rotation held that falling bodies would be left behind as the earth turned beneath them; hence they should appear to land to the west if the earth rotates. Newton's experiment hoped to show that they land instead to the east. The tangential velocity of the top of a high tower is greater than that of the foot; therefore a falling body should

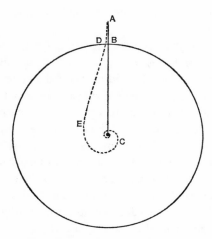

Figure 6. Newton's sketch of the path of a
body that falls on a rotating earth.

slightly outrun the place directly beneath its point of release. A master
experimenter, Newton carefully defined the details of the trial to ensure
its accuracy. He also drew a trajectory that showed the path as part of a
spiral ending at the center of the earth. (See Figure 6.)

He concluded the letter by saying that he would certainly comment on
Hooke's hypothesis if he had seen it and would with pleasure hear objections against any of his own notions. "But yet my affection to Philosophy
being worn out, so that I am almost as little concerned about it as one
tradesman uses to be about another man's trade or a country man about
learning, I must acknowledge my self avers from spending that time in
writing about it wch I think I can spend otherwise more to my own content
& ye good of others . . . ."

The spiral was a gross blunder. By drawing the complete curve as
though the earth were not present to offer resistance, Newton implicitly
converted the problem of fall into the problem of orbital motion and
showed a body with an initial tangential velocity falling into the center of
attraction. Hooke can be excused for correcting the mistake. He expressed his conviction that under conditions of no resistance, a body let
fall on a rotating earth would not fall to the center but would rather follow

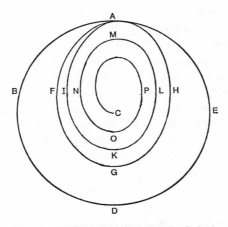

Figure 7. AFGH is the elliptical path of a body that meets no resistance. The inner spiral AIKLMNOPC represents its path in a resisting medium.

forever a path resembling an ellipse. (See Figure 7.) He explicitly treated the problem as orbital motion by referring it to "my Theory of Circular motions compounded by a Direct motion and an attractive one to a Center."

When he received the correction, Newton's announced pleasure in hearing objections evaporated like a dew in August. He had been caught, as Hooke had been caught in 1672, by haste. We can judge how deep an impression Hooke's correction made by the fact that six-and-a-half years later he remembered the correspondence in detail. More than thirty years later, the memory still smarted enough that he tried to explain the error away by calling it "a negligent stroke with his pen," which Hooke interpreted as a spiral. At the time, he knew that the diagram was not a negligent stroke of the pen, however, and he knew as well that he had in his letter explicitly called it a spiral. His reply, as he accepted Hooke's correction, was as dry as a piece of burned bacon. Yes, Hooke was correct; the body would not descend to the center "but circulate wth an alternate ascent & descent . . . ." Hooke was mistaken, however, about the ellipse. In a brief but compelling argument, Newton showed that under uniform gravity the body would reach its lowest point in a little less than 120 degrees and its original height in a little less than 240, and he closed by leaving his newest offering to Hooke's correction. He got it.

> Your Calculation [Hooke replied] of the Curve by a body attracted by an aequall power at all Distances from the center Such as that of a ball Rouling in an inverted Concave Cone is right and the two auges [apsides] will not unite by about a third of a Revolution. But my supposition is that the Attraction always is in a duplicate proportion to the Distance from the Center Reciprocall, and Consequently that the Velocity will be in a subduplicate proportion to the Attraction and Consequently as Kepler Supposes Reciprocall to the Distance.

The passage has drawn as much attention as the hint at universal gravitation in his lecture about the system of the world. On close analysis, its apparent derivation of the inverse-square law turns out to be a bastard demonstration resting on a deep confusion about dynamics and accelerated motion. Newton understood as much and chose not to answer a third time.

Newton later acknowledged that he did take up the challenge in the exchange of letters and demonstrated (in a paper which survives) that an elliptical orbit around an attracting body located at one focus entails an inverse-square attraction. If the inverse-square relation had initially flowed from the substitution of Kepler's third law into the formula for centrifugal force under the simplifying assumption of circular orbits, the demonstration of its necessity in elliptical orbits far excelled in difficulty what had been a simple substitution. In fact, the demonstration, which probably dated from early 1680, was one of the two foundation stones on which the concept of universal gravitation rested. Newton did not even consider sending it to Hooke. Neither, in 1680, did he pursue it any further himself. He had worn out his affection for philosophy and now devoted his time to other studies more to his own content.

Nevertheless, the correspondence with Hooke had wider ramifications that bore on philosophy. Hooke posed the problem of orbital motion in terms of an attraction to the center, an action at a distance similar to the short-range attractions and repulsions Newton seemed ready to embrace in 1679. Both in his letters to Hooke and in his demonstration with the ellipse, Newton accepted the concept of attraction without even blinking. Between "De aere et aethere," which I have placed in 1679, and papers from 1686–7, Newton did not compose any essay that we know of on the system of nature. We have no way to determine if he adopted in 1679–80 opinions later expressed, and if so, whether Hooke's proposal served to crystallize opinions beginning to take form, or whether he found Hooke's concept of attraction acceptable because they had taken form already. What we do know is that Newton had recast his entire philosophy of nature by 1686–7. In papers composed with the *Principia*, first a proposed "Conclusio," then the same material in a proposed "Preface," neither of which appeared in the published work, Newton applied action at a distance to virtually all the phenomena of nature. The "Conclusio" started by noting that in addition to the observable motions in the cosmos there were innumerable unobservable ones among the particles of bodies.

If any one shall have the good fortune to discover all these, I might almost say that he will have laid bare the whole nature of bodies so far as the mechanical causes of things are concerned. I have least of all undertaken the improvement of this part of philosophy. I may say briefly, however, that nature is exceedingly simple and conformable to herself. Whatever reasoning holds for greater motions, should hold for lesser ones as well. The former depend upon the greater attractive forces of larger bodies, and I suspect that the latter depend upon the lesser forces, as yet unobserved, of insensible particles. For, from the forces of gravity, of magnetism and of electricity it is manifest that there are various kinds of natural forces, and that there may be still more kinds is not to be rashly denied. It is very well known that greater bodies act mutually upon each other by those forces, and I do not clearly see why lesser ones should not act on one another by similar forces.

In the evidence he proceeded to offer in support of his assertion, Newton composed an early draft of what later became familiar as Query 31 of the *Opticks*. All of the crucial phenomena that had caught his attention in 1664–5 and had played roles connected with aetherial mechanisms in the "Hypothesis of Light" appeared. All now furnished evidence of forces of attraction and repulsion between particles. The overwhelming burden of the argument rested, however, on chemical phenomena, as it continued to do in Query 31. Newton drew on many examples that fell into two main categories: reactions that generate heat and reactions that reveal affinities. The first repeated the argument he made to Boyle in 1679, with attractions now supplying the new motion manifest as heat. The latter converted the secret principle of sociability into specific attractions between certain substances. Without exception, all the chemical phenomena he cited had appeared in his alchemical papers, some among his own experiments, some among the other papers, most of them in both. Certain alchemical themes, such as the role of fermentation and vegetation in altering substances, the peculiar activity of sulfur, and the combination of active with passive, also made their way into the new exposition of the nature of things.

Newton understood that he was proposing a radical revision of what he referred to as "philosophy of the common sort . . . ." Although the very words *attractions* and *repulsions* would displease many, he wrote in one draft,

yet what we have thus far said about these forces will appear less contrary to reason if one considers that the parts of bodies certainly do cohere, and that distant particles can be impelled towards one another by the same causes by which they cohere, and that I do not define the manner of attraction, but speaking in ordinary terms call all forces attractive by which bodies are impelled towards each other, come together and cohere, whatever the causes be.

The last statement seemed to open the door again to the aetherial mechanisms of philosophy of the common sort. Recall that it was written to appear at the conclusion of a treatise which contained a pendulum experiment designed to disprove the existence of an aether and which also contained in its Book II a sustained argument against the Cartesian philosophy in particular and mechanical philosophies in general.

As it appears to me, Newton's philosophy of nature underwent a profound conversion in 1679–80 under the combined influence of alchemy and the cosmic problem of orbital mechanics, two unlikely partners which made common cause on the issue of action at a distance. Insofar as he continued to speak of particles of matter in motion, Newton remained a mechanical philosopher in some sense. Henceforth, the ultimate agent of nature would be for him a force acting between particles rather than a moving particle itself – what has been called a dynamic mechanical philosophy in contrast to a kinetic. In the realm of chemical phenomena, he never succeeded in moving the concept of force beyond the level of general speculation. His paper of 1680, however, showed the possibilities inherent in the idea when it was applied on the cosmic plane and supported by all the resources of Newton's mathematics. In 1680, the problem failed to seize his imagination, however, and he put it aside.

૨▲

Hooke's letter was not the last. During the winter of 1680–1, a comet appeared, and the comet entailed a new intrusion. As all astronomers in

Europe with one exception believed, not one comet but two appeared. The first was sighted before sunrise early in November and vanished into the morning sun at the end of the month. Two weeks later, in mid-December, another comet appeared in the early evening, moving away from the sun. In late December, the second comet was immense, with a tail four times as broad as the moon and more than seventy degrees long. "I beleive scarce a larger hath ever been seen . . . ," an excited Royal Astronomer, John Flamsteed, wrote to a friend in Cambridge. Flamsteed was the astronomer who believed that the two comets were one and the same, a single comet which reversed its direction in the vicinity of the sun, a notion which marked a radical break with universally accepted opinion.

Flamsteed knew where to turn for comment. Although he had met Newton in Cambridge in 1674, he had never corresponded with him, and he approached him now through a friend, James Crompton, a fellow of Jesus College. Meanwhile, the comet had kindled Newton's own curiosity. On 12 December, only four days after its first evening sighting, he observed it and recorded information about its tail. He observed it almost daily from that time until it disappeared in March, keeping a log on its tail.

The majesty of the heavens had imprinted itself indelibly on Newton's imagination in Grantham and would soon dominate him again. It held him for a time in 1681. In addition to observing, he systematically collected the observations of others, and he began to read what literature he could find: Hooke, Hevelius, Gottignies, and Petit. He worked at reducing the observations to a path through space. He also wrote two long letters to Flamsteed criticizing his theory. The length of the letters, in contrast to his reticence with Hooke, and the fact that he addressed the second directly to Flamsteed instead of Crompton, give some measure of the extent of his involvement.

Newton refused to accept Flamsteed's new theory. According to the theory, the comet did not circle the sun; it turned in front of it. Manifold valid objections sprang to Newton's mind, and he laid them out at some length. The primary interest of the letter lies in what it did not contain, however. Only a year before, Newton had solved the mechanics of orbital motion for a planet circling the sun. He did not now attempt to apply the same principles to the comet. Hence the letter allows us to measure crudely the progress of his thought toward the concept of universal gravitation. Most opinion held that comets were foreign bodies not related to the solar system and not governed by its laws. In his writings on comets,

Hooke excluded them from the attraction between cosmic bodies that he posited. Apparently Halley held a similar view in 1680. The letter to Flamsteed strongly implies that Newton did also. That is, no matter how important his demonstration was that elliptical orbits entail an inverse-square force, Newton apparently considered the force as specific to the solar system, which contained related bodies. He had not yet formulated the idea of universal gravitation.

Although Newton did not continue the correspondence with Flamsteed, his interest in comets did not evaporate. When the comet of 1682, which we now call Halley's comet, appeared, he made and recorded observations of its position. Some time after 1680, he systematically collected information on all recorded comets and classified them under a number of headings, such as those in opposition to the sun. He also revised his opinion on the paths of comets. In a set of propositions about comets, along with assertions that the sun and planets have gravitation toward their centers that decreases as the square of the distance and that the sun's gravitation is far greater than the planets', he forsook the theory of rectilinear paths for comets and accepted curved ones. The point of greatest curvature coincides with the point of perihelion. If the comet returns, the curve is an "oval"; if it does not return, it is nearly a hyperbola. Whatever his reluctance in the spring of 1681, Newton now saw the application to comets of the orbital dynamics of planets.

Newton's mathematical renown beyond the university's walls also impinged upon his privacy. In 1683, John Wallis published *A Proposal about Printing a Treatise of Algebra*, which announced that he would expound Newton's method of infinite series. When his *Treatise of Algebra* did appear in 1685, it devoted five chapters to portions of the two *Epistolae* to Leibniz of 1676. More important, both for its testimony and for its impact, was a letter from Edinburgh, which he received in June 1684, from David Gregory, the nephew of James Gregory.

S^r

Altho I have not ye honour of your acquaintance, yet ye character and place yee bear in the learned world I presume gives me a title to adress you especially in a matter of this sort, which is humbly to present you with a treatise latly published by me heir, which I am sure contains things new to the greatest parte of the geometers.

The work that the letter accompanied was *Exercitatio geometrica de dimensione figurarum* (*Geometrical Exercise on the Measurement of Figures*), an exposition of his uncle's method of infinite series with applications to quadratures, volumes, and the like. Gregory told Newton that he knew from Collins's letters to his uncle that Newton had cultivated this method and that the world had long expected to see his discoveries.

S^r yee will exceedingly oblidge me, if yee will spare so much time from your Philosophical and Geometrical studies, as to allow me your free thoughts and character of this exercitation, which I assure you I will justly value more then that of all the rest of ye world . . . .

Gregory's letter stirred Newton to action. Because Gregory had handsomely acknowledged Newton in the work, no charge of plagiarism could arise. In its threat of forestalling Newton, however, Gregory's book was like Mercator's earlier one, and he responded in the same way. He projected a treatise in six chapters with the title "Matheseos universalis specimina" ("Specimens of a Universal System of Mathematics"), in which he intended to include letters that would demonstrate his priority over James Gregory. From the beginning, he also had Leibniz in mind, and he indicated his intention to publish their entire correspondence of 1676–7. In the chapters he wrote, Newton virtually forgot Gregory and devoted himself to replying to Leibniz. Hence the "Matheseos" provides a sharp insight into Newton. Despite his declared lack of interest, he had not been able to put the German completely out of his mind. Gregory posed no threat. Newton recognized that Leibniz did, and a defensive polemical tone characterized the treatise as he answered objections now seven years old. The projected six chapters concentrated on infinite series. The chapter 4 that he wrote expounded the method of fluxions. It began by translating the fluxional anagram in the *Epistola posterior,* and it ended by comparing the fluxional method with Leibniz's differential calculus as he had stated it in the letter of 1677. Though Leibniz's definitive publication of his calculus had not yet begun, he had published in 1682 a quadrature of the circle which Newton later cited against him. Perhaps Newton was aware of it as he composed the "Matheseos." More likely he was responding to the threat he had perceived in 1676–7.

In chapter 4, Newton also repeated a recent attack on modern analysts that appears among his mathematical papers, Leibniz among them by implication. One fascinating sentence suggests the depth his revulsion against Descartes, which was central to his attack on analysts, had reached. After expounding upon his fluxional method, he paused to reflect. "On these matters I pondered nineteen years ago, comparing the findings of and Hudde with each other." The silence of the blank is deafening. Only one name – Descartes – could have belonged there. Newton could no longer bring himself even to acknowledge his debt.

Newton left the "Matheseos" unfinished and started a revision with the title "De computo serierum" ("On the Computation of Series"). He tired of it rather quickly as well and abandoned it in the middle of the third chapter. He never returned to either.

On the Continent, someone else had noted the publication of Gregory's work. In July, Otto Menke, editor of the *Acta eruditorum*, wrote to Leibniz that someone in England had attributed his quadrature of the circle to Newton. Leibniz had already clashed with another German mathematician, Ehrenfried von Tschirnhaus, about papers on tangents and maxima and minima, which Leibniz called plagiarism. Menke's message spurred him to compose a paper on his differential calculus, which the *Acta* published in October. In the summer of 1684, Newton and Leibniz seemed destined to collide, as indeed they were. The collision was deferred, however, largely because of yet another interruption. In August, Edmond Halley traveled up from London to put a question that only Newton could answer. No other interruption stirred him so deeply. Halley's visit changed the course of his life.

8

Principia

THE BACKGROUND to Halley's visit to Cambridge in August 1684 was a chance conversation of the previous January. By his own account, Halley had been contemplating celestial mechanics. From Kepler's third law, he had concluded that the centripetal force toward the sun must decrease in proportion to the square of the distance of the planets from the sun. The context of his statement implied that he arrived at the inverse-square relation by substituting Kepler's third law into Huygens's recently published formula for centrifugal force. He was not the only one who made the substitution. After Hooke raised the cry of plagiarism in 1686, Newton recalled a conversation with Sir Christopher Wren in 1677 in which they had considered the problem "of Determining the Hevenly motions upon philosophicall principles." He had realized that Wren had also arrived at the inverse-square law. It is clear that the problem Hooke put to Newton in the winter of 1679–80 was one that several people defined for themselves at much the same time. It was, indeed, the great unanswered question confronting natural philosophy, the derivation of Kepler's laws of planetary motion from principles of dynamics.

This same problem was discussed by Halley, Wren, and Hooke at a meeting of the Royal Society in January 1684. Hooke claimed that he could demonstrate all the laws of celestial motion from the inverse-square relation. Halley admitted that his own attempt to do so had failed. Wren was skeptical of Hooke's claim. Hooke again asserted that he had the demonstration, but he intended to keep it secret until others, by failing to solve the problem, learned how to value it. We do not know what took Halley to Cambridge. Because he allowed seven months to pass, we can hardly surmise that he rushed there, afire with curiosity, to lay the problem before Newton. Nevertheless, he did find himself in Cambridge in August, and he did seize the opportunity to consult a man he knew to be expert in mathematics.

Although Halley mentioned the visit, the best account came from Newton's recollection as he told it to Abraham DeMoivre.

In 1684 Dr Halley came to visit him at Cambridge, after they had been some time together, the D^r asked him what he thought the Curve would be that would be described by the Planets supposing the force of attraction towards the Sun to be reciprocal to the square of their distance from it. Sr Isaac replied immediately that it would be an Ellipsis, the Doctor struck with joy & amazement asked him how he knew it, why saith he I have calculated it, whereupon Dr Halley asked him for his calculation without any farther delay, Sr Isaac looked among his papers but could not find it, but he promised him to renew it, & then to send it him . . . .

We can dismiss the charade of the lost paper – all the more because it survives among Newton's papers. Newton did not lightly send things abroad. The repeated faux pas in the correspondence with Hooke on this very question would have made him more wary than usual. He did commit himself to reexamining the paper, however, by promising to send the demonstration to Halley.

As Newton told the story to DeMoivre, he must have congratulated himself for restraining any impetuosity. When he tried the demonstration anew, it did not work out. As he ultimately discovered, a hastily drawn diagram had led him to confuse the axes of the ellipse with conjugate diameters. Not one to give up, he started anew and finally achieved his goal. In November, via the hands of Edward Paget, Halley received somewhat more than he expected, a small treatise of nine pages with the title *De motu corporum in gyrum* (*On the Motion of Bodies in an Orbit*). Not only did it demonstrate that an elliptical orbit entails an inverse-square force to one focus, but it also sketched a demonstration of the original problem: An inverse-square force entails a conic orbit, which is an ellipse for velocities below a certain limit. Starting from postulated principles of dynamics, the treatise demonstrated Kepler's second and third laws as well. It hinted at a general science of dynamics by further deriving the trajectory of a projectile through a resisting medium. When he received *De motu*, Halley did not wait another seven months. He recognized that the treatise embodied a step forward in celestial mechanics so immense as to constitute a revolution. Without delay, he made a second trip to Cambridge to confer with Newton about it; and on 10 December, he reported his activities to the Royal Society.

Mr. Halley gave an account, that he had lately seen Mr. Newton at Cambridge, who had shewed him a curious treatise, *De motu;* which, upon Mr. Halley's desire, was, he said, promised to be sent to the Society to be entered upon their register.

Mr. Halley was desired to put Mr. Newton in mind of his promise for the securing his invention to himself till such time as he could be at leisure to publish it. Mr. Paget was desired to join with Mr. Halley.

Letters from Newton to Flamsteed indicate that Halley was not the only one to sense the revolutionary implications of the treatise. His copy was in such demand that Flamsteed, to whom Newton offered the privilege of reading it, had to wait a month before he could see it.

The cause of Halley's long wait for the amended treatise was a process at work in Cambridge, a process typically Newtonian though not less marvelous for that. The problem had seized Newton and would not let him go. There was in it that same majesty that years before had aroused the awe of a schoolboy in Grantham. Over the years, he had briefly heard its call several times. As a student, he had found the inverse-square relation from Kepler's third law. Under Hooke's stimulus, he had extended the inverse-square force to account for Kepler's first law. In August 1684, Halley evoked the same splendor anew, and this time Newton surrendered utterly to its allure. Halley later liked to say that he had been "the Ulysses who produced this Achilles," but Halley in London did not understand what was happening in Cambridge. Halley did not extract the *Principia* from a reluctant Newton. He merely raised a question at a time when Newton was receptive to it. It grasped Newton as nothing had before, and he was powerless in its grip. The treatise *De motu*, which Halley received in November, bore marks testifying that the challenge was at work already. Initially the treatise contained four theorems and five problems that dealt with motion in space void of resistance. As he began to glimpse broader horizons, Newton revised his initial definitions and hypotheses to allow for resistance and added two problems on motion through such a medium. At much the time when Halley was telling the Royal Society about *De motu*, Newton was writing to Flamsteed for data that would enable him to make its demonstrations more precise. "Now I am upon this subject," he told Flamsteed in January, "I would gladly know ye bottom of it before I publish my papers." In getting to the bottom of it, he nearly cut himself off from human society. From August 1684 until the spring of 1686, his life is a virtual blank except for the *Principia*. In December and January 1684–5, he wrote to Flamsteed four times for information, and again in September he asked for more. In February 1685, he responded to a letter about *De motu* from the secretary of the Royal Society, Francis Aston, his old acquaintance in Trinity. In April, Newton heeded the appeal of William Briggs and wrote a letter of muted praise which Briggs prefixed to the Latin translation of his *New Theory of Vision*. He allowed family obligations to draw him to Woolsthorpe briefly in the spring of 1685. Beyond this meager list of known activities, most of which were related to the *Principia*, there was nothing. The investigation wholly absorbed his life. Apparently he continued to lecture. William Whiston, who entered Clare in 1686, remembered that he heard one or two of Newton's lectures, which he did not understand. The manuscripts that Newton later deposited as lectures, however, were merely drafts of the *Principia*. He forwent alchemical experimentation, which had been a major activity since 1678.

An awestruck Humphrey Newton observed the erratic behavior of a man transported outside himself.

So intent, so serious upon his Studies, y^t he eat very sparingly, nay, ofttimes he has forget to eat at all, so y^t going into his Chamber, I have found his Mess untouch'd of wch when I have reminded him, [he] would reply, Have I; & then making to ye Table, would eat a bit or two standing . . . At some seldom Times when he design'd to dine in ye Hall, would turn to ye left hand, & go out into ye street, where making a stop, when he found his Mistake, would hastily turn back, & then sometimes instead of going into ye Hall, would return to his Chamber again. . . . When he has sometimes taken a Turn or two [in the garden], has made a sudden stand, turn'd himself about, run up ye Stairs, like another Alchimedes [*sic*], with an εὕρηκα, fall to write on his Desk standing, without giving himself the Leasure to draw a Chair to sit down in.

Early in February, external events threatened to shatter his concentration. Charles II died, passing the crown on to his brother and heir, James II, an acknowledged Catholic. On the morning of 9 February the university gathered at the public schools, all attired in their robes, and proceeded to Market Hill to proclaim James. No sooner had Newton retired to his room than the mayor and aldermen of the city, mounted and attired in their equally resplendent robes and accompanied by the councilmen, bailiffs, and freemen of Cambridge, appeared before the gate of Trinity, immediately outside his window, to proclaim James once more. Two years elapsed before the crisis that the succession implied touched Cambridge. Fortunately, two years were exactly the time he required.

The *Principia* was not only Newton's monumental achievement. It was also the turning point in his life. As we know from his papers, he had performed prodigies in a number of fields. As we also know, he had completed nothing. By 1684, he had littered his study with unfinished mathematical treatises. He had not pursued his promising insights in mechanics. His alchemical investigations had produced only a chaos of unorganized notes and disconnected essays. Had Newton died in 1684 and his papers survived, we would know from them that a genius had lived. Instead of hailing him as a figure who shaped the modern intellect, however, we would at most mention him in brief paragraphs lamenting his failure to reach fulfillment. The period 1684–7 brought an end to the tentative years. Finally, he saw an undertaking through to its conclusion. True, as the goal came in sight, leaving the excitement of discovery behind and the drudgery of computation ahead, he began once again to lose interest and to procrastinate. But this time the sheer grandeur of the theme carried him through to completion. The publication of the *Principia* could not reshape Newton's personality, of course, but the magnitude of its achievement thrust him into the public eye beyond the possibility of another withdrawal.

The *Principia* redirected Newton's intellectual life, which theology and alchemy had dominated for more than a decade. It interrupted his theological studies, which he did not resume seriously for another twenty years. It did not terminate his career as an alchemist, but it diverted the thrust of alchemical concepts from the private world of arcane imagery into an unexpected and concrete realm of thought where the rigor of mathematical precision could help them reshape natural philosophy. The investigation that seized Newton's imagination late in 1684 and dominated it for the following two-and-a-half years transformed his life as much as it transformed the course of Western science.

❧

When he started, Newton did not realize where the pursuit would lead or the demands it would place on him. What he sent to Halley was a short treatise concerned primarily with orbital mechanics. It implied that inverse-square centripetal attractions are general in nature because it asserted both that the satellites of Jupiter and Saturn, as well as the planets about the sun, conform to Kepler's third law and that the motions of comets are governed by the same laws that determine planetary orbits. It was about these things that Newton wrote to Flamsteed. He wanted Flamsteed's observations of the periods of Jupiter's satellites and the dimensions of their orbits. Flamsteed indicated that they indeed also obey Kepler's third law. "Your information about ye Satellits of Jupiter gives me very much satisfaction," Newton assured him. Newton also asked for the precise celestial coordinates of two stars in the constellation Perseus. As seen from the earth, the great comet of 1680–1 had passed through Perseus. Flamsteed understood immediately that Newton had taken the comet under consideration again. Newton confirmed his surmise. "I do intend to determin ye lines described by ye Comets of 1664 & 1680 according to ye principles of motion observed by ye Planets . . . ."

De motu had already included the matter above. When Newton asked Flamsteed for observations of the velocities of Jupiter and Saturn as they approach conjunction, however, he was raising a question which *De motu* had not addressed. Every time Jupiter approaches conjunction with Saturn, Saturn ought to slow down in its orbit and then to speed up as Jupiter passes conjunction "(by reason of Jupiters action upon him) . . . ." The mutual influence of Saturn and Jupiter implied the existence of a universal attraction. Flamsteed was skeptical about such notions. When they are nearest each other, Jupiter and Saturn are separated by a distance four times the radius of the earth's orbit. The aether, which is a yielding matter, would simply absorb any disturbance at that distance. Nevertheless, he did supply exactly the information that Newton wanted to hear about the motion of Saturn and Jupiter in conjunction. Every new question broadened the investigation. When he replied to Aston in February, Newton apologized for his delay in sending the emended treatise. He had intended to complete it sooner, "but the examining severall thinges has taken a greater part of my time then I expected, and a great deale of it to no purpose."

When he wrote to Flamsteed again in September to obtain more data about the comet of 1680–1, he had yet another question. Flamsteed had published some observations about tides in the estuary of the Thames, and Newton wanted additional information about them, indicating to us if not to Flamsteed a significant further extension of the gravitational concept. Against the background of a constantly expanding study, we can appreciate the significance of Newton's comment to Halley in 1686 about a paper of the 1660s in which he tried to calculate how much the centrifugal force due to the earth's rotation decreases gravity. "But yet to do this business right is a thing of far greater difficulty then I was aware of." The greater difficulty measured the difference between *De motu* and the ultimate *Principia.*

The first serious problem that Newton confronted was dynamics itself. If the seventeenth century had taken giant strides in mechanics, it had yet to crown its efforts with a science of dynamics. In order to write the *Principia,* Newton had first to create a dynamics equal to the task, and he devoted much of the six months that followed the composition of *De motu* to this task.

De motu started with two definitions and two hypotheses. Definition 1 drew upon the lesson about circular motion that Hooke had taught him in 1679 to introduce a new word into the vocabulary of mechanics.

I call that by which a body is impelled or attracted toward some point which is regarded as a center centripetal force.

Newton later explained that he coined the word *centripetal,* seeking the center, in conscious parallel to Huygens's word *centrifugal,* fleeing the center. No single word better characterized the *Principia,* which more than anything else was an investigation of centripetal forces as they determine orbital motion.

Definition 2 concerned rectilinear motion.

And [I call] that by which it endeavors to persevere in its motion in a right line the force of a body or the force inherent in a body.

Hypothesis 2 extended the definition into a general conception of motion.

By its inherent force alone, every body proceeds uniformly in a right line to infinity unless something extrinsic hinders it.

The combined statement of the definition and the hypothesis is a startling assertion to find in the first draft of the work that established the principle of inertia as the foundation of modern science. Together they indicate the extent of the task Newton faced in autumn 1684.

Three versions of *De motu* survive. The second version was merely a fair copy in Halley's hand of the first, with small additions that Newton had indicated his intention to insert. The third version, on the other hand, contained the beginning of Newton's reconstruction of his dynamics. To the two original hypotheses, modestly altered, he added three more and then rejected the name "Hypothesis" in favor of a new one, "Lex." Hence Newton's laws of motion initially numbered five. From its new position at the head of the list, Law 1 continued to assert that a body moves uniformly by its inherent force alone. In Law 2, Newton attempted to define the action of impressed force.

The change of motion is proportional to the impressed force and is made in the direction of the right line in which that force is impressed.

The crux of Newton's dynamics lay in the relation of inherent force and impressed force, what he later called (as he struggled to clarify them) "the inherent, innate and essential force of a body" and "the force brought to bear or impressed on a body." The continuing development of his dynamics hinged on the two concepts.

The third version of *De motu* contained another feature, a scholium to Problem 5 on "the immense and truly immobile space of the heavens." Like the concept of the inherent force of bodies, which he held to be the distinguishing characteristic of true motion, the scholium looked back to Newton's revulsion from the relativism of Cartesian physics. In his earlier paper "De gravitatione," he had said that the ultimate absurdity of Cartesian relativism lay in its consequence "that a moving body has no determinate velocity and no definite line in which it moves." In its original form, *De motu* did not contain any reference to absolute space. It had no need to. The inherent force of bodies adequately defined their absolute motions. As the demands of an internally consistent dynamics swept Newton inexorably toward the principle of inertia, he began to insist on absolute space. In papers that revised the third version, he considerably expanded its brief statement as he moved still further toward the principle of inertia. The increasing insistence of his assertions looked forward to the well-known scholium on absolute space and time which he included in the *Principia.* He may have capitulated to Descartes on motion. On the issue of relativism, which in Newton's view smacked of atheism, he continued to shout his defiance until his dying day.

In two papers of revisions that followed the third version of *De motu,*

Newton completed the transformation of his dynamics into its ultimate form. By now, he had set his mind on logical rigor. Where the third version had defined four terms, the first paper of revisions defined no fewer than eighteen at one stage or another. Many concerned themselves with absolute space and absolute motion. Accompanying them, alterations in the definitions of motion transformed his concept of inherent force as he moved toward a quantitatively rigorous dynamics.

The inherent, innate, and essential force of a body is the power by which it perseveres in its state of resting or of moving uniformly in a right line, and is proportional to the quantity of the body. It is actually exerted proportionally to the change of state, and in so far as it is exerted, it can be called the exerted force of a body . . . .

In the second paper of revisions, Newton introduced a further change in the definition of inherent force whereby he assigned it, not to a body, but to matter, as he did in the *Principia*. Later he suggested another name, *vis inertiae*, the force of inertia. He revised Law 1 in a similar way to state that a body, by its inherent force alone, perseveres in its state of resting or of moving uniformly in a straight line. That is, inherent force ceased to be the cause of uniform motion. An addition to his definition of impressed force ratified the change. "This force consists in the action only, and remains no longer in the body when the action is completed." With the alterations in the definitions and in Law 1, Newton effectively embraced the principle of inertia. In the *Principia* itself, he further eliminated the reference to inherent force from the statement of the first law, thus obliterating the principal record of the path by which he arrived at it.

Once he adopted the principle of inertia, the rest of his dynamics fell quickly into place. He had seized on the essence of his second law twenty years before and had never altered it as he wrestled with the first law. Impressed force alters a body's motion; the change of motion is proportional to it. In this proportionality lay the possibility of a quantitative science of dynamics that would cap and complete Galileo's kinematics.

The statement of the second law implied a quantity which he now defined, quantity of motion. It required in turn a further definition, quantity of matter. The quantity of a body, he said, "is calculated from the bulk of the corporeal matter which is usually proportional to its weight." When he revised the definitions, he both separated this one and placed it at the head of the list.

The quantity of matter is that which arises conjointly from its density and magnitude. A body twice as dense in double the space is quadruple in quantity. This quantity I designate by the name of body or of mass.

Without the concept of mass, here defined satisfactorily for the first time, the second law, the force law, would have remained incomplete. The one required the other. Together they constituted the heart of Newton's contribution to dynamics.

The concept of mass included more than mere quantity of matter, however. As he revised his first law of motion toward the principle of inertia, he transferred his revised notion of inherent force from that law, where it had become an embarrassment, to the concept of mass.

The Inherent force of matter is the power of resisting by which any body, as much as in it lies, perseveres in its state of resting or of moving uniformly in a right line: and it is proportional to its body and does not differ at all from the inactivity of the mass except in our mode of conceiving it. In fact a body exerts this force only in a change of its state effected by another force impressed upon it, and its Exercise is *Resistance* and *Impetus* which are distinct only in relation to each other.

In the classic phrase of the seventeenth century, matter is indifferent to motion. Leibniz was to argue that if matter is wholly indifferent to motion, any force should be able to impart any velocity to a body, and a quantitative science of dynamics is impossible. Although Newton offered no rationale, it is evident that he responded to similar considerations. As an activity of resistance evoked in changes of state, mass establishes the equation between an impressed force and the change of motion it produces.

Thus his final dynamics continued to focus on the interplay of inherent force and impressed force, with the former wholly transformed from its original form. To replace the parallelogram of forces, which related the two in the original version of *De motu*, he devised a further law of motion which has come down to us in a different wording as the third law.

As much as any body acts on another so much does it experience in reaction. . . . In fact this law follows from Definitions 12 [the inherent force of a body to persevere in its state of resting or moving uniformly] and 14 [the force brought to bear and impressed on a body to change its state] in so far as the force of the body exerted to conserve its state is the same as the force impressed on the other body to change its state, and the change of state of the first body is proportional to the first force and of the second body to the second force.

We cannot date with precision the sheets that definitively revised the dynamics of *De motu*, but they appear to have belonged to the early months of 1685. Few periods have held greater consequences for the history of Western science than the three to six months in the autumn and winter of 1684–5, when Newton created the modern science of dynamics. Whereas the invisible mechanisms of orthodox mechanical philosophy, such as Descartes's vortices, had continually diverted attention away from quantitative precision toward picturable images, Newton's new conception of action at a distance invited mathematical treatment. The first stage of Newton's work on the *Principia* was the creation of his dynamics, the instrument that the rest of the task required.

ᴥ

One other development of note took place in the early months. A scholium in the third version of *De motu* that contained a rough correlation of the moon's orbit with the acceleration of gravity on the earth referred to media whose density (or quantity of solid matter) is "almost proportional to their weight . . . ." He made a similar comment in his paper of revisions that followed the third version. The quantity of a body is calculated "from the bulk of the corporeal matter which is usually proportional to its weight." Almost proportional, usually proportional – however broadly Newton's thoughts were ranging, he had not yet arrived at his final conception of universal gravitation if he could use such phrases. In the second case, he even included a practical device to compare the solid matter in two bodies of equal weight: Hang them from equal pendulums; the quantity of matter varies inversely as the number of oscillations made in the same time. Newton crossed the passage out. On a blank verso opposite, he wrote that the weight of heavy bodies is proportional to their quantity of matter as experiments with pendulums can prove: "When experiments were carefully made with gold, silver, lead, glass, sand, common salt, water, wood, and wheat, however, they resulted always in the same number of Oscillations."

Now at last, the full implications of an idea only half explored before could open before him. The equal acceleration in free fall of all heavy bodies found its explanation; the pendulums offered a demonstration of the same phenomenon which was far more delicate, as tiny differences would accumulate on successive swings and become manifest. In the heavens, Kepler's third law did the same, unless one accepted the unlikely assumption that the planets are exactly equal in mass. The satellites of Jupiter were so many more celestial pendulums. Not only did their conformance to Kepler's third law reveal the proportionality of their masses to their attractions toward Jupiter, but their concentric orbits about Jupiter demonstrated that the sun attracts both them and Jupiter in proportion to mass. According to the third law, the satellites of Jupiter must attract the sun in return. Apparently every body in the world attracts every other body. We can only imagine the wild surmise that raced through Newton's mind as the principle of universal gravitation silently unfolded. As he pursued orbital mechanics, his new quantitative dynamics had led him to a generalization more inclusive than any that natural philosophy had hitherto proclaimed. The published *Principia* repeatedly insisted that the mathematical treatment of attraction asserted nothing about its physical cause. When he first wrote his discovery down, he was less reserved; the forces proportional to the quantity of matter, he said, "arise from the universal nature of matter . . . ." He appears to have arrived at the concept very nearly at the time when his namesake, Alderman Samuel Newton, proclaimed James Stuart king of England immediately outside his window.

Clearly, a discovery so grandiose required an exposition commensurate to it. The dynamics he had created provided an adequate instrument. Newton began to expand *De motu* into a systematic demonstration of universal gravitation. We know very little about the process except that the nine-page tract of November 1684 had expanded to a treatise in two books of well over ten times that length by roughly the following November. To it he gave the name *De motu corporum* (*On the Motion of Bodies*).

Although we do not know at what point he attacked it, we do know that the attraction of a sphere early loomed as a major problem. The correlation of the moon's orbit with the acceleration of gravity assumed that the inverse square law holds, not only at the distance of the moon but also at the surface of the earth. All the particles of the vast earth, stretching out in every direction beyond the horizon, combine to attract an apple a few feet above its surface at Woolsthorpe or at Cambridge with a force dependent, not on the apple's distance from the surface, but on its distance from the center of the earth. Was such a notion credible? As Newton told it to Halley in 1686, at least, it was not.

I never extended yᵉ duplicate proportion lower then to yᵉ superficies of yᵉ earth & before a certain demonstration I found yᵉ last year have suspected it did not reach accurately enough down so low . . . There is so strong an objection against yᵉ accurateness of this proportion, yᵗ without my Demonstrations . . . it cannot be beleived by a judicious Philosopher to be any where accurate.

Hence the significance of Proposition XL of *De motu corporum* (corresponding to proposition LXXI, Book I), in which Newton demonstrated that a homogeneous spherical shell, composed of particles that attract inversely as the square of the distance, attracts a particle external to it, no matter at what distance, inversely as the square of its distance from the center of the sphere.

As Newton realized, the correlation between the moon and the apple ought to be more precise than "very nearly," the words he used in the third version of *De motu*. The calculation he now carried out showed a correlation correct to the last inch (or one part in four hundred) of the acceleration of gravity as Huygens had determined it.

With the demonstration of a sphere's attraction, and with the exact correlation of the moon's motion with the measured acceleration of gravity, the logical foundation of the concept of universal gravitation became secure. Newton was a man of rigor. Though his imagination ranged widely, he could not consider putting so vast an idea as universal gravitation in print before he had satisfied himself of its demonstrability. So far, he had proved the presence of inverse-square attractions in the solar system, indeed the presence of a single inverse-square attraction. Moreover, the earth must attract the moon to hold it in its orbit; and if apples fall to the earth, it must attract them as well. By what right could he extend the ancient word *gravitas* (heaviness), applied to the apple, both to the attraction that holds the moon and to the attraction of the sun? Only the inverse-square correlation of the moon with the apple, plus a liberal dash of the principle of uniformity, permitted the argument. Together with Proposition XI (that elliptical orbits entail inverse-square attractions), the demonstration of a sphere's attraction was one of the two foundation stones on which the law of universal gravitation rested.

❧

The plan of the initial *De motu corporum* began, after the definitions, laws of motion, and exposition of the mathematical method of ultimate ratios, in which he expressed the work, with a somewhat enlarged set of the orbital propositions that had formed the heart of the preceding *De motu*. So far, the exposition had dealt with the abstract problem of bodies in motion about unspecified centers of force, which to Newton always meant physical bodies. According to the third law, central attracting bodies must themselves be attracted and moved. *De motu corporum* now turned to consider the complications that mutual attractions introduce. First, Newton dealt with the problem of two mutually attracting bodies. The solar system consists of many bodies, however: a sun and six planets (as far as the seventeenth century knew), three of which (again as far as they knew) have satellites. Could so many mutual attractions fail to upset the orbital dynamics demonstrated for single bodies? The many-body problem presented a more serious challenge than the two-body problem, and one that Newton could not avoid. He had taken as his province the explanation of natural philosophy from the principle of attraction. Abstract demonstrations, however elegant, were one thing. Natural philosophy addressed itself to the real world, and the real world consisted of many bodies in motion, all of them, on Newton's hypothesis, attracting one another. Newton found that a demonstrative solution to the problem exceeded his power. (Indeed, it can now be shown to be impossible.) Nevertheless, he did find an analytic tool by which to attack the simplest form of the problem, three mutually attracting bodies, the conclusions from which he could extend by plausible arguments to a system of many bodies.

The analysis, in Proposition XXXV (Proposition LXVI, Book I), considered the case of a large central body, *S* (standing for *Sol*, the sun) circled by two planets *P* and *Q*. He asked himself what perturbations the attraction of the outer planet *Q* introduces into the motion of the inner planet *P*. To answer the question, he analyzed the attraction of *Q* on *P* into two components, a radial one, *LM*, and one that is more perturbing, *MQ*. (See Figure 8.) Because *Q* attracts both *P* and *S*, the disturbing effect of the component *MQ* is only that portion (*MN*) by which it differs from the accelerative attraction of *Q* on *S* (*NQ*). If several planets, *P*, *Q*, *R*, etc., revolve around a great central body *S*, he concluded, "the motion of the innermost revolving body *P* is least disturbed by the attractions of the others, when the great body is attracted and agitated by the others equally in proportion to weights and distances and they by each other."

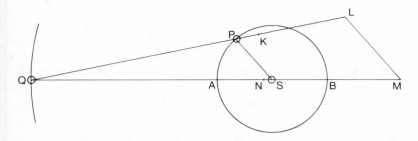

Figure 8. No diagram such as this survives. It is reconstructed from the text by analogy with the diagram to Proposition LXVI.

And hence if several smaller bodies revolve about the great one, it may easily be inferred that the orbits described will approach nearer to ellipses, and the descriptions of areas will be more nearly uniform, if all the bodies mutually attract and agitate each other equally in proportion to their weights and distances, and if the focus of each orbit is placed in the common center of gravity of all the interior bodies . . . than if the innermost body were at rest, and were made the common focus of all the orbits.

Originally, the analysis of the three-body problem directed itself solely to the problem of multiple planets around the sun and the apparent conflict between universal gravitation and Kepler's laws. Almost at once, Newton saw that it could be turned to quite the opposite use, not to explaining how orbits can so closely approximate Kepler's laws despite mutual attraction but to explaining the source of observed perturbations. First of all, this meant the moon. By the time Newton drafted Book II of *De motu corporum,* he had composed a new proposition directed to systems, like the earth and moon, that orbit central attracting bodies together. To the analysis of the forces in play he added twenty-one additional corollaries to supplement the original one. The majority addressed themselves to the inequalities of lunar motion, but the analysis could extend beyond lunar inequalities. Let the orbit of *P* represent a continuous ring of particles instead of the orbit of a single body, and let its radius be equal to the radius of the earth. When the ring of particles is a fluid, the analysis deals with the tides. Newton composed Corollaries 18 and 19 about them. When the ring is solid and attached to the earth, the analysis offers a derivation of precession. Corollaries 20 and 21 concerned themselves with that phenomenon.

In contrast to the starkly mathematical format of Book I, the second book of the expanded work presented a prose essay on the Newtonian system of the world based on the propositions demonstrated in Book I and referring to them for support. We can date the time of its composition with some confidence to the autumn of 1685. From the observed phenomena of the heliocentric system, Newton argued the necessity of attractive forces to hold bodies in closed orbits, of inverse-square attractive forces to sustain orbits stable in space and systems that conform to Kepler's third law, and finally of a single inverse-square attractive force that arises "from the universal nature of matter." Most of Book I had looked directly to that conclusion.

Book II proceeded then briefly to suggest, on the basis of Proposition XXXVI, that the principle of universal gravitation could account for other observed phenomena. One paragraph asserted that it could explain the observed anomalies of the moon's motion, a second that it predicted further anomalies as yet unobserved. He devoted single paragraphs, devoid of quantitative details as the paragraphs on the moon were, to precession and tides and turned to a long qualitative discussion of comets which led to the conclusion that comets come under the laws of orbital motion developed for planets. He was not able, however, to produce a method that could successfully determine the orbit of a specific comet.

In fact, Book II never existed in the form described above. In the very process of composing it, Newton introduced a change pregnant with further possibilities. As soon as Humphrey had copied the paragraph that dealt with tides, before he could proceed to the next paragraph, Newton crossed it out, and in its place he began to compose a detailed, quantitative discussion of tides based on Proposition XXXVI and its twenty-one new corollaries. He extended the original paragraph to an eighteen-paragraph discussion which has all the elements of the one that finally appeared in the *Principia*. Here, as far as the original manuscript of the final book is concerned, he stopped for the moment.

❧

The two books, *De motu corporum,* probably completed in autumn 1685, were not yet the *Principia*. The excitement of the pursuit still held him;

and as further horizons continued to open before him, he allowed them to lead him on. What he had written so far was an argument, containing a few digressions, for the principle of universal gravitation. He now began to expand it further into an investigation of motions terrestrial and celestial which would of course still conclude in the law of universal gravitation but would also present a general system of dynamics, propose on its basis a new ideal of science, and demonstrate the impossibility of the rival Cartesian system.

As Newton generalized his dynamics, he recognized the privileged status of two force laws: forces that decrease in proportion to the distance squared, and forces that increase in proportion to the distance. They alone are compatible with elliptical orbits. In their cases alone spheres composed of attracting particles attract by the same law as the particles. He discovered still another remarkable analogy in his investigation of the stability of orbits. Among physically probable laws of attraction, only these two laws yield orbits whose lines of apsides do not move. "When I wrote my treatise about our Systeme," Newton wrote to the Reverend Richard Bentley a few years later, "I had an eye upon such Principles as might work wth considering men for the beleife of a Deity . . . ." From the composition of the work, it appears more likely that he fastened his eye on the most important principle only in the process of writing it. Only two laws of attraction, the inverse-square law and the direct-distance law, are compatible with a rationally ordered universe. God showed himself to be Newton's equal in mechanics when He built His cosmos upon them.

As far as the cosmos was concerned, Newton did not consider the law of attraction in proportion to distance a likely candidate. In an infinite universe, it entails the impossible consequences of infinite forces, infinite accelerations, and infinite velocities. It is, moreover, empirical fact that attractions decrease with distance rather than increase – further testimony to God's good sense. The law of force in proportion to distance plays a role in natural philosophy, however, primarily in vibratory motions like those of pendulums.

Although Section IX, on the stability and motion of the apsides of orbits, contributed to the generalization of Newton's dynamics, it concerned itself primarily with a specific problem, the moon's orbit. Its conception belonged to the same period in which Newton added the expanded body of corollaries to Proposition XXXVI (Proposition LXVI,

Book I) and began to contemplate an exact quantitative account, not merely of the gross phenomena of the cosmos but also of the fine deviations from ideal patterns as well. The perturbations of the moon offered the prime target for such a program; and the progression of the moon's line of apsides, in contrast to the virtual stability of planetary orbits, presented the first problem. Section IX offered a sophisticated analysis of the effect that variations from an inverse-square force have on the stability and motion of apsides. Specifically, he derived the quantitative effect that the presence of the sun should have on the moon's orbit. He must have viewed the result both as a victory and as a defeat. He had brought a prominent feature of the moon's motion, hitherto recognized as an observed fact which defied explanation, within the scope of his celestial dynamics. The mere fact that he published the analysis indicates that he regarded the result as a considerable achievement. Alas, the quantity that he obtained was only half the observed progression of the lunar apsides. We can measure the extent to which Newton felt the defeat by the fact that he could not bring himself to acknowledge the discrepancy. Only in the third edition, after he had abandoned hope of correcting it, did he insert a brief statement that the progression of the apse is about twice as swift.

According to his letter to Halley in the summer of 1686, Newton completed his expansion of Book I during the winter of 1685–6. The same letter indicated that he also added to the twenty-two propositions that dealt with motions in and phenomena of resisting media as he expanded Book I. When it began to reach excessive size, he decided to divide the book into two, Book I to concern itself entirely with the motions of bodies in spaces free of resistance, and Book II with various problems of resisting media. Deeming Book I complete, he had Humphrey copy it and dispatched the manuscript to the Royal Society. Regarding his task as nearly finished, he relaxed and resumed alchemical experimentation that spring after he had ignored his laboratory during the whole of 1685. In a famous memorandum written some thirty years later in connection with the calculus controversy, Newton asserted, "The Book of Principles was writ in about 17 or 18 months, whereof about two months were taken up with journeys, & the MS was sent to ye R. S. in spring 1686; & the shortness of the time in which I wrote it, makes me not ashamed of having committing some faults." Perhaps Newton did later regard spring 1686 as the point at which he completed the *Principia*. The last clause indicates that he had an interest in claiming as short a time as possible, however, and too much other evidence points to substantial further work on Books II and III.

Already he had set about revising the manuscript of what would now be Book III. An emendation entered a reference to Robert Hooke in the second paragraph. Originally the text had said merely that recent philosophers postulate either vortices or some other principle of impulse or attraction to explain how planets are drawn back from rectilinear paths and constrained to move in closed orbits. Newton altered the statement to read: "More recent philosophers postulate either vortices, as Kepler and Descartes do, or some other principle of impulse or attraction, as in the case of Borelli, Hooke, and others of our nation." In the context of the *Principia*, which also postulated a principle of attraction for that purpose, the addition can be read only as a generous acknowledgment of his debt to Hooke.

ॐ

On 21 April 1686, Halley told the Royal Society that Newton's treatise was nearly ready for the press. A week later, Newton gave Halley's promise concrete support.

Dr. Vincent presented to the Society [the minutes record] a manuscript treatise entitled *Philosophiae Naturalis principia mathematica,* and dedicated to the Society by Mr. Isaac Newton, wherein he gives a mathematical demonstration of the Copernican hypothesis as proposed by Kepler, and makes out all the phaenomena of the celestial motions by the only supposition of a gravitation towards the center of the sun decreasing as the squares of the distances therefrom reciprocally.

It was ordered, that a letter of thanks be written to Mr. Newton; and that the printing of his book be referred to the consideration of the council; and that in the meantime the book be put into the hands of Mr. Halley, to make a report thereof to the council.

We can attribute the description of the content of the manuscript to the fact that Halley had recently been appointed clerk to the society and hence wrote the minutes himself.

Three weeks passed. Nothing happened. The Royal Society was in disarray at the time, frequently not meeting for lack of an officer present to preside. No council convened to consider Newton's manuscript. Increasingly anxious that Newton have some response, Halley apparently seized the initiative and raised the question at the society's meeting on 19 May. Although such matters fell within the sole competence of the council, the society voted – nay, ordered –

That Mr. Newton's *Philosophiae naturalis principia mathematica* be printed forthwith in quarto in a fair letter; and that a letter be written to him to signify the Society's resolution, and to desire his opinion as to the print, volume, cuts, &c.

If the resolution was indeed Halley's work, he took a considerable risk in promoting it. Though reared in a wealthy family, Halley had found himself reduced to relative penury by the death of his father in 1684. With a young wife and a family and no means of livelihood, he had accepted the humble post of clerk to the Royal Society at a salary of £50 per annum. Menial staff are not ordinarily expected to initiate projects of major import, neither in the Royal Society of the seventeenth century nor elsewhere. Before the year was out, Halley's audacity had nearly cost him his job.

Meanwhile, the vote freed him to write officially to Newton, telling him, in terms of glowing praise, of the society's action. He also informed Newton that he was to be in charge of the publication, and he got down to business at once by urging that the diagrams be enlarged. Halley was the more anxious to write before Newton should hear of something else through other channels.

There is one thing more that I ought to informe you of, viz, that Mr Hook has some pretensions upon the invention of ye rule of the decrease of Gravity, being reciprocally as the squares of the distances from the Center. He sais you had the notion from him, though he owns the Demonstration of the Curves generated therby to be wholly your own; how much of this is so, you know best, as likewise what you have to do in this matter, only Mr Hook seems to expect you should make some mention of him, in the preface, which, it is possible, you may see reason to praefix. I must beg your pardon that it is I, that send you this account, but I thought it my duty to let you know it, that so you may act accordingly; being in myself fully satisfied, that nothing but the greatest Candour imaginable, is to be expected from a person, who of all men has the least need to borrow reputation.

The first paragraph of Halley's letter contained all the praise any man might need to receive. The second paragraph was another matter, and characteristically, Newton's reply focused exclusively on it. "I thank you for w^t you write concerning Mr Hook," he began, "for I desire that a good understanding may be kept between us." Even across the space of three centuries, one can hear Halley's sigh of relief as he read the letter. Annoyed perhaps but well short of anger, Newton recited the events of 1679. The expected crisis apparently past, Halley sent a proof of the first sheet for Newton to approve the type. He did not neglect to offer some additional flattery to soothe any feelings still ruffled. He had proofread the sheet, he told Newton, but he might have missed some errors; "when it has past your eye, I doubt not but it will be clear from errata."

Halley relaxed too soon. Newton's anger required time to ripen fully. For three weeks he fed on Hooke's charge while his fury mounted. Late in June, he was finally ready to write again, a display of pyrotechnics worthy of the ardent young man who in 1672 had cast restraint to the winds rather than swallow Hooke's condescension.

In order to let you know yᵉ case between Mʳ Hook & me [he announced without further ado to a startled Halley, who had hoped to hear no more about it], I gave you an account of wᵗ past between us in our Letters so far as I could remember. . . . I intended in this Letter to let you understand yᵉ case fully but it being a frivilous business, I shal content my self to give you yᵉ heads of it in short: vizt yᵗ I never extended yᵉ duplicate proportion lower then to yᵉ superficies of yᵉ earth & before a certain demonstration I found yᵉ last year have suspected it did not reach accurately enough down so low: & therefore in yᵉ doctrine of projectiles never used it nor considered yᵉ motion of yᵉ heavens: & consequently Mʳ Hook could not from my Letters wᶜʰ were about Projectiles & yᵉ regions descending hence to yᵉ center conclude me ignorant of yᵉ Theory of yᵉ Heavens. That what he told me of yᵉ duplicate proportion was erroneous, namely that it reacht down from hence to yᵉ center of yᵉ earth. That it is not candid to require me now to confess my self in print then ignorant of yᵉ duplicate proportion in yᵉ heavens for no other reason but because he had told it me in the case of projectiles & so upon mistaken grounds accused me of that ignorance. That in my answer to his first letter I refused his correspondence, told him I had laid Philosophy aside, sent him only yᵉ experimᵗ of Projectiles (rather shortly hinted then carefully described) . . . to sweeten my Answer, expected to heare no further from him, could scarce perswade my self to answer his second letter, did not answer his third, was upon other things, thought no further of philosophical matters then his letters put me upon it, & therefore may be allowed not to have had my thoughts of that kind about me so well at that time.

As he continued his litany of complaint, Newton made it clear that he had spent some time reviewing his papers. To support his prior knowledge of the inverse-square relation, he cited his early paper on the endeavor of the planets and the moon to recede, his letter to Oldenburg when he received Huygens's *Horologium,* and the implication of his explanation of gravity in the "Hypothesis of Light." Even if he granted that he got the inverse-square law from Hooke, Hooke only guessed at it, whereas he had demonstrated its truth. So much for the past. He had designed the whole treatise to consist of three books. The second was short and required only copying. The third he had decided to suppress. In a final cry of exasperation, all the pent-up tension of a year and a half of stupendous and unremitting toil burst out. "Philosophy is such an impertinently litigious Lady that a man had as good be engaged in Law suits as have to do with her. I found it so formerly & now I no sooner come near her again but she gives me warning." Before he could mail the letter, he received a report, undoubtedly via Paget, that Hooke was making a stir and demanding that justice be done. Even more enraged now, he added a postscript longer than the letter in which he recited his complaint once more for Halley's edification. All Hooke had done was to publish Borelli's hypothesis under his own name, and now he claimed to have done everything but the drudgery of calculation.

Now is not this very fine? Mathematicians that find out, settle & do all the business must content themselves with being nothing but dry calculators & drudges & another that does nothing but pretend & grasp at all things must carry away all the invention as well of those that were to follow him as of those that went before.

Hooke had stated nothing that any mathematician could not have told him once Huygens's work had been published, and in extending the duplicate proportion to the center of the earth, he had fallen into error.

And why should I record a man for an Invention who founds his claim upon an error therein & on that score gives me trouble? He imagins he obliged me by telling me his Theory, but I thought my self disobliged by being upon his own mistake corrected magisterially & taught a Theory wch every body knew & I had a truer notion of then himself. Should a man who thinks himself knowing, & loves to shew it in correcting & instructing others, come to you when you are busy, & notwithstanding your excuse, press discourses upon you & through his own mistakes correct you & multiply discourses & then make this use of it, to boast that he taught you all he spake & oblige you to acknowledge it & cry out injury & injustice if you do not, I beleive you would think him a man of a strange unsociable temper. Mr Hooks letters in several respects abounded too much wth that humour wch Hevelius & others complain of . . . .

Apparently nothing galled him so much as the demand for acknowledgment. He returned to it in the letter three times. Acknowledge Hooke's priority? Quite the contrary, he went back to the draft of the final book and attacked the reference he had made to Hooke. He slashed out the acknowledgment of Hooke's concept of attraction in the second paragraph. Later on, the discussion of comets had included an observation by "Cl [Clarissimus] Hookius"; one brutal stroke of the pen reduced "the very distinguished Hooke" to mere "Hooke." When he recast the book, he went a step further and eliminated the passage altogether, as he did another one that acknowledged an observation by Hooke.

In trying circumstances, Halley proved himself a diplomat of no small capacity. His letter indicates that he did not fully understand what Newton's threat to suppress Book III meant. Recalling the two-book manuscript he had seen, he thought Newton intended merely to separate the theory of comets from the system of the world. Whatever he understood, he did not intend to stand by silent as Newton castrated his masterwork. He cajoled. He flattered. He assured Newton that the Royal Society stood on his side. He urged him not to suppress Book III, "there being nothing which you can have compiled therin, which the learned world will not be concerned to have concealed . . . ." He took care to follow Newton's suggestion that he question Wren about a conversation in 1677, and he reported on it wholly in Newton's favor. He recounted his meeting with Hooke and Wren in January 1684, again to Newton's advantage. He gave a fuller account of the events at the meeting when Dr. Vincent presented the manuscript, assuring him that Hooke's behavior had been represented to Newton as worse than it was. True, when the members of the society forgathered as usual at a coffee house after the meeting, Hooke claimed he had given the invention to Newton, "but I found that they were all of opinion, that nothing therof appearing in print, nor on the Books of the Society, you ought to be considered as the Inventor . . . what application he has made in private I know not, but I am sure that the Society have a very great satisfaction in the honour you do them, by your dedication of so worthy a Treatise." And finally, Halley did not hesitate to give Newton some straight talk.

Sʳ I must now again beg you, not to let your resentments run so high, as to deprive us of your third book, wherein the application of your Mathematicall doctrine to the Theory of Comets, and severall curious Experiments, which, as I guess by what you write, ought to compose it, will undoubtedly render it acceptable to those that will call themselves philosophers without Mathematicks, which are by much the greater number.

Calmed by such ministrations, the storm blew itself out. When he wrote again two weeks later, Newton regretted that he had added the postscript to the earlier letter and even admitted his debt to Hooke in three things. To compose the dispute, he enclosed a revision of the scholium to Proposition IV. One might search for some time to find a less generous acknowledgment. Proposition IV stated the formula for centripetal force in circular motion, and Corollary VI showed that an inverse-square force follows when periods and times correspond to Kepler's third law. The scholium pointed out that the case of Corollary VI obtains in the celestial bodies. To the statement he added a parenthetical comment: "(as our countrymen Wren, Halley & Hooke have also severally concluded)." Not only did he put Hooke's name last, but he added two additional paragraphs, intended both for Hooke's edification and for his own justification. Ever the diplomat, Halley quietly altered the order of names so that Hooke's appeared before his own in the published work.

With one more blast at the end of July, Newton left Halley in peace. Because he did not explicitly withdraw his threat to suppress Book III, he also left Halley in doubt as to what else he would receive.

❧

Meanwhile, the *Principia* proceeded. Though Newton did not utter threats lightly, the work had established over him a command that he could not break. The comparison of 1672 with 1686 is instructive. At the first word of criticism of his paper on colors, Newton began to withdraw into his shell and ended up by severing his connections with the learned world in London. The provocation in 1686 was far worse; the result was quite opposite. If Halley recognized the monumental importance of the manuscript, if the Royal Society greeted it in a similar way, if Hooke called it the most important discovery in nature since the world's creation and tried to claim it as his own, much more did its author understand the significance of his own work. He could make threats in anger. He could not mutilate his masterpiece. The charge of plagiarism and the reply to it provided at most an interlude, a momentary release from the tension of composition. The interlude completed, he resumed composition again as though it had never happened.

There was also a new factor in the situation. He had a publisher. The arrival of Newton's manuscript in April 1686 confronted the Royal Society with an unprecedented problem. A year and a half before, it had urged Newton to send his treatise *De motu* to be entered on its register in order to secure the invention to him until he was ready to publish. The society itself was not a publisher, however, and it had not intended to function as such. Unexpectedly, it found itself in receipt of a manuscript of obvious importance and dedicated to it, at a time when it was virtually bankrupt. On 19 May, Halley led the society to order that the *Principia* be printed. Finances belonged to the province of the council, however; and when the council met on 2 June, it pleasantly decided to let Halley stew in the broth he had mixed. "It was ordered, that Mr. Newton's book be printed, and that Mr. Halley undertake the business of looking after it, and printing it at his own charge; which he engaged to do." One can understand the dimension of personal anguish in Halley's successful attempt to calm Newton's rage against Hooke. In effect, he was the publisher.

By Newton's own testimony, he completed Book II in its final form during the autumn of 1686. In expanding it into its published form, Newton converted it into an attack on Descartes. Sections VI and VII explored the physical cause of resistance, in order to demonstrate that the Cartesian plenum was flatly incompatible with the observed phenomena of nature. In case the theory of resistance were not enough, Newton closed the General Scholium on resistance at the end of Section VI with an account of his experiment to disprove that an aether exists.

With serious wounds already inflicted on Cartesian natural philosophy, Section IX delivered the coup de grâce by attacking that part of Cartesian philosophy most relevant to the *Principia*'s central concern, the theory of vortices. In keeping with the tone of the work as a whole, he delivered the attack by means of a mathematical analysis of the dynamic conditions of vortical motion, something no Cartesian had undertaken. (See Figure 9.) Two damning conclusions emerged from the analysis. First, a vortex cannot sustain itself. Its continuation in a steady state requires the constant transfer of motion from layer to layer until, in Newton's words, it is "swallowed up and lost" in the boundlessness of space. Therefore, a vortex without a continuing source of new energy – or, in Newton's terms, new motion – must decay. In the second place, the periods of revolution vary in a vortex as the square of the radius, whereas Kepler's third law,

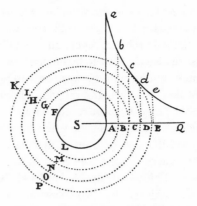

Figure 9. The dynamics of a vortex.

based on the celestial phenomena, demands the three-halves power. "Let philosophers then see," Newton concluded, "how that phenomenon of the 3/2th power can be accounted for by vortices."

On the surface, the problem posed to philosophers does not appear so difficult, because the whole analysis depended on an arbitrary assumption about the friction between infinitesimal layers of the vortex. The problem that Newton set the vortical theory was more profound, however. A scholium that concluded both Section IX and Book II formulated it in terms of the velocities in a single orbit. The essence of the dilemma lies in the incompatibility between the velocity relations in Kepler's second law and those in his third. No assumption about friction can remove it. One and the same vortex cannot correspond to the different variations in velocity required by Kepler's second and third laws, "so that the hypothesis of vortices is utterly irreconcilable with astronomical phenomena," Newton concluded, "and rather serves to perplex than explain the heavenly motions. How these motions are performed in free spaces without vortices, may be understood by the first Book; and I shall now more fully treat of it in the following Book."

❧

Meanwhile, as Newton worked on Book II, Halley contracted with a printer, established the book's style (including such things as woodcuts of diagrams, which could then appear on the same page with the text), and set publication in motion. Almost at once, Newton's irrepressible doubts and hesitations began to emerge. For no obvious reason, he requested that the work not appear before the end of the Michaelmas term. With the Hooke imbroglio on his hands, Halley had no desire to cross the wishes of his author. Hence the project moved forward slowly, and only thirteen sheets had been printed by October. The press then came to a halt for four months; in February, Newton had not received anything to proofread beyond the eleventh sheet. Ignorance about Newton's intentions and Halley's assumption from what he knew that the rest would be short probably contributed to the halt. After the spate of letters about Hooke, Halley heard almost nothing. He visited Cambridge near the first of September. He must have learned no more than he heard by mail, for in the winter he had no idea what else he might expect to receive from Newton.

During the same period when he was expanding Book II, Newton was also revising the final book; indeed, he was completely recasting it.

In the preceding books [he began, in his new introduction] I have laid down the principles of philosophy; principles not philosophical but mathematical: such, namely, as we may build our reasonings upon in philosophical inquiries. These principles are the laws and conditions of certain motions, and powers or forces, which chiefly have respect to philosophy. . . . It remains that, from the same principles, I now demonstrate the frame of the System of the World.

He insisted on the word *mathematical.* He had originally composed the final book in a popular form so that many could read it;

but afterwards, considering that such as had not sufficiently entered into the principles could not easily discern the strength of the consequences, nor lay aside the prejudices to which they had been many years accustomed, therefore, to prevent the disputes which might be raised upon such accounts, I chose to reduce the substance of this Book into the form of Propositions (in the mathematical way), which should be read by those only who had first made themselves masters of the principles established in the preceding Books.

Years later, Newton recited a similar story to his friend William Derham. He abhorred contentions, he said. "And for this reason, namely to avoid being baited by little Smatterers in Mathematicks, he told me, he designedly made his Principia abstruse . . . ."

It has been almost universally accepted that the recasting of Book III, from the prose essay that constituted the original Book II to the mathe-

matical format that he published, flowed from the clash with Hooke.
Hooke was the little smatterer in mathematics; Newton would show him
what was what – and for that matter, who was who – by composing Book
III, the climax of the work, in a form Hooke could not even follow. Such
an account hardly squares with the facts. True, Newton did recast the
format in a mathematical style, so that Book III proceeded with the
paraphernalia of propositions and lemmas. As far as the material found in
the original Book II is concerned, however, the change was purely cos-
metic. Through Proposition XVIII (together with some of the later propositions) the final Book III did not differ substantially from the earlier draft.
The new Book III began with Proposition XIX, which derived the ratio
between the axis of the earth and its diameter at the equator. It included a
vastly expanded lunar theory, in which Newton demonstrated quantitatively various inequalities in the moon, and a derivation of the precession of the equinoxes which was also quantitative. In a word, it proposed a
new ideal of a quantitative science, based on the principle of attraction,
which would account not only for the gross phenomena of nature but also
for the minor deviations of the gross phenomena from their ideal patterns.
Against the background of inherited natural philosophy, this was a conception no less revolutionary than the idea of universal gravitation itself.
Newton had begun to glimpse its possibility in the winter of 1685–6. It
animated the twenty-one new lemmas added to Proposition LXVI, the
extended discussion of tides (which went into Book III with no significant
addition), the new paragraph about the moon, and Section IX, Book I, on
the stability and motion of apsides. Newton made all of these alterations
before the episode with Hooke. Because he was at his irritable worst
when the excitement of discovery drew him taut, it may well be that the
relation between the explosion that disintegrated Hooke and the recasting
of Book III was the exact opposite of the accepted account. That is, the
tension of the revision may have caused the explosion. I thus find it
impossible to believe that Newton's threat to suppress Book III, uttered at
the very time when the wider vision of the book was opening itself to his
view, was ever more than a passing expression of exasperation.

The revised Book III contained one other thing. Comets finally yielded
to his assault. When he wrote to Halley in the summer of 1686, he had
not yet succeeded with them. Sometime during the following nine
months, he did. He later told Gregory "that this discussion about comets

is the most difficult in the whole book." It was more difficult than most of us realize: from observations made on the earth as it moves in an elliptical orbit to determine the conical path of a comet moving in a different plane. Needing a concrete case to make his theory compelling, he succeeded in determining that the comet of 1680–1 had moved in a plane inclined at an angle of 61° 20 1/3′ to the ecliptic. By locating its perihelion, he fixed the axis of its orbit, and he calculated its latus rectum. The comet traced a conic (which he treated as a parabola in edition one) describing areas in proportion of time. That is, the laws of planetary mechanics based on the attraction of the sun also governed the motions of the comet of 1680–1 and by inference the motions of all comets.

As the *Principia* became more radical in the new ideal of science that it proposed, Newton began to worry about the reception of the concept of attraction. Initially, he had intended to state his position straightforwardly. In the early versions of Books I and II, he had spoken of attractions without apology and described the gravity of cosmic bodies as a force that arises from the universal nature of matter. He had drafted a Conclusio, similar to what he later published as Query 31 of the *Opticks*, in which he suggested the existence of a wide range of other forces between the particles of matter. So far, he began, he had treated the system of the world and the great motions in it that are readily observed.

There are however innumerable other local motions which on account of the minuteness of the moving particles cannot be detected, such as the motions of the particles in hot bodies, in fermenting bodies, in putrescent bodies, in growing bodies, in the organs of sensation and so forth. If any one shall have the good fortune to discover all these, I might almost say that he will have laid bare the whole nature of bodies so far as the mechanical causes of things are concerned.

By analogy with macroscopic motions and the gravitational forces that control them, he urged that similar forces between particles cause the microscopic motions. As we have seen elsewhere, he proceeded then to cite the phenomena, first of all chemical phenomena that he had observed in his laboratory, then the other crucial phenomena that had figured in his speculations for twenty years which seemed to demand a restructuring of natural philosophy on the concept of forces. He gave up the Conclusio and included the same material in the draft of a preface. In the end, he suppressed both and in the ultimate preface referred only to his belief in "certain forces by which the particles of bodies, by some causes hitherto unknown, are either mutually impelled towards one another, and cohere in regular figures, or are repelled and recede from one another."

Newton had good reason to be cautious. Weaned on the mechanical philosophy himself, he could not doubt how the concept of a universal attraction of all particles of matter for one another would be received. Thus he began to hedge. The line in the preface about bodies being impelled by causes unknown was only part of a more extensive camouflage in which he insisted that his mathematical demonstrations did not entail any assertions about the ontological status of forces. He amended the second paragraph of original Book II with such a statement. As he wrote it first, it concluded by asserting the necessity of some force, which he called in general centripetal force, that draws bodies back from their rectilinear paths and holds them in closed orbits. Now he added a further statement.

But our purpose is only to trace out the quantity and properties of this force from the phenomena, and to apply what we discover in some simple cases as principles, by which, in a mathematical way, we may estimate the effects thereof in more involved cases . . . We said, *in a mathematical way*, to avoid all questions about the nature or quality of this force, which we would not be understood to determine by any hypothesis . . . .

He inserted similar disclaimers at the beginning and end of Section XI. He could have saved his breath. Fifty such disclaimers – or fifty times fifty – would not have stifled the outrage of mechanical philosophers.

≥●

Newton apparently finished Book II sometime during the winter of 1686–7. Meanwhile, as we have seen, the process of publication had come to a halt. Whatever Newton's contribution, the primary cause was a crisis in the Royal Society that touched Halley. On 29 November, the council suddenly resolved that Halley's continuation in his office should be put to a vote and that a new election of a clerk in his place should be held. On 5 January following, the council appointed a committee to investigate his performance of his duties. We know nothing about the background of this attack. On the surface, at least, it looks political. Although nothing that even approaches definitive evidence exists, there are some grounds for believing that Hooke's circle of supporters mounted the attack on Halley, perhaps for his leading role in the society's endorsement of the *Principia*, though it is necessary to add that Halley's correspondence gave no hint that Hooke conducted such a vendetta.

Whether such was the case or not, the council's committee reported in Halley's favor on 9 February. It could have been no later than the following day when he wrote to Newton for a copy of a sheet of the manuscript lost by the printer and indicated his intention to push the publication toward completion. In his reply on 13 February, Newton thanked Halley for "putting forward ye press again . . . ." The second book was complete. Halley could have it whenever he wished.

Newton's letter, together with the following correspondence, is interesting for the light it sheds on Halley's relation to him. Until he received it early in March, Halley did not know what Book II would contain. Because Newton had not responded to his plea the previous summer, he did not know if there was to be a Book III, let alone its contents. Only with that information can we appreciate the magnitude of the gamble Halley took. And only by appreciating the gamble can we appreciate the impact of Newton's manuscript on him. Though the first, he was not the last to experience its power.

With the information that Book II was on its way, Halley engaged a second printer to set it while the first completed Book I. Apparently he expected the short manuscript Newton had promised the previous summer. When he received it early in March, he found a treatise nearly as long as the first book and filled, like the first, with new investigations of which he had never dreamed. Holding his breath, he finally chanced an oblique question about the matter Newton had left in doubt the previous June.

You mention in this second, your third Book *de Systemate mundi*, which from such firm principles, as in the preceding you have laid down, cannot chuse but give universall satisfaction; if this be likewise ready, and not too long to be got printed by the same time, and you think fit to send it; I will endeavour by a third hand to get it all done togather, being resolved to engage upon no other business till such time as all is done: desiring herby to clear my self from all imputations of negligence, in a business, wherin I am much rejoyced to be any wais concerned in handing to the world that that all future ages will admire.

In fact, Book III was virtually ready, and only a month later, Halley wrote to assure Newton that he had received the third part of his "divine

Treatise. . . ." Because the first printer had nearly completed Book I, he did not have to employ a third but gave Book III to him.

For Halley, a hellish four months ensued. The eighteenth sheet of Book I, mostly concerned with the difficult Proposition XLV, on the motion of apsides, gave him "extraordinary trouble," and he feared they might need to reset it. In April, he excused himself for not writing to Wallis by his attention to Newton's book; "the correction of the press costs me a great deal of time and paines." In June, he pled the same excuse for the same lapse. He even felt compelled to insert an apologetic "Advertisement" at the end of issue number 186 of the *Philosophical Transactions,* immediately after his review of the *Principia,* explaining why the issue was appearing three months late. For his devoted labor, which absorbed most of his time for a year, Halley received a handsome acknowledgment in the preface. As far as we know, it never occurred to Newton to thank him as well. Halley told both Newton and Wallis that he hoped to complete the edition by the Trinity term, that is, by 21 June. As it worked out he missed by two weeks. On 5 July, he announced that the task was finally completed.

Honoured Sr

I have at length brought your Book to an end, and hope it will please you. the last errata came just in time to be inserted. I will present from you the books you desire to the R. Society, Mr Boyle, Mr Pagit, Mr Flamsteed and if there be any elce in town that you design to gratifie that way; and I have sent you to bestow on your friends in the University 20 Copies, which I entreat you to accept.

Newton sent Humphrey about Cambridge presenting the twenty copies to his acquaintances and to the heads of the colleges, "some of wch (particularly Dr. Babington of Trinity) said that they might study seven years, before they understood anything of it." A student had the last word. As Newton passed him in the street, he delivered the ultimate benediction of Restoration Cambridge on the genius it harbored: "there goes the man that writt a book that neither he nor any body else understands."

9

Revolution

NEWTON was hardly an unknown man in philosophic circles before 1687. The very extent to which he had made his capacity in physics and mathematics known had functioned in the early 1680s to destroy his attempt to reconstruct an isolation in which he might pursue his own interests in his own way. Nevertheless, nothing had prepared the world of natural philosophy for the *Principia*. The growing astonishment of Edmond Halley as he read successive versions of the work repeated itself innumerable times in single installments. Almost from the moment of its publication, even those who refused to accept its central concept of action at a distance recognized the *Principia* as an epoch-making book. A turning point for Newton, who, after twenty years of abandoned investigations, had finally followed an undertaking to completion, the *Principia* also became a turning point for natural philosophy. It was impossible that Newton's life could return to its former course.

Rumors of the coming masterpiece had flowed through Britain during the first half of 1687. For those who had not heard, a long review in the *Philosophical Transactions* announced the *Principia* shortly before publication. Although the review was unsigned, we know that Halley wrote it. With the exception of Newton himself, no one knew the contents of the work better. He insisted on its epochal significance.

This incomparable Author [the review began] having at length been prevailed upon to appear in publick, has in this Treatise given a most notable instance of the extent of the powers of the Mind; and has at once shewn what are the Principles of Natural Philosophy, and so far derived from them their consequences, that he seems to have exhausted his Argument, and left little to be done by those that shall succeed him.

After the body of the review presented a summary of the *Principia*, Halley closed with a further encomium: "[I]t may be justly said, that so many and so Valuable *Philosophical Truths*, as are herein discovered and put past Dispute, were never yet owing to the Capacity and Industry of any one Man."

In mathematical circles, such as the one gathered around David Gregory in Scotland, the fame and influence of the *Principia* spread quickly. Across the Channel, a political refugee, John Locke, set himself to mastering the book. Because he was not a mathematician, he found the demonstrations impenetrable. Not to be denied, he asked Christiaan Huygens if he could trust the mathematical propositions. When Huygens assured him he could, he applied himself to the prose and digested the physics without the mathematics. Locke realized that Newton was one of the intellectual giants of the age. He made it his business to meet Newton upon his return to England, and he included an admiring reference to him in the preface to his *Essay on Human Understanding* (1690).

In London, young Abraham DeMoivre met the *Principia* by chance when he happened to be at the home of the Duke of Devonshire at the time (probably in 1688) when Newton came to present a copy. DeMoivre was supporting himself by teaching mathematics. All of twenty-one years old, he considered himself the perfect master of the subject.

The young mathematician opened the book and deceived by its apparent simplicity persuaded himself that he was going to understand it without difficulty. But he was surprised to find it beyond the range of his knowledge and to see himself obliged to admit that what he had taken for mathematics was merely the beginning of a long and difficult course that he had yet to undertake. He purchased the book, however; and since the lessons he had to give forced him to travel about continually, he tore out the pages in order to carry them in his pocket and to study them during his free time.

Like Gregory, he succeeded ultimately in enrolling himself among the disciples of his new master.

Newton's book was acknowledged and recognized on the Continent as well as in Britain. During the spring and summer of 1688, three of the leading continental journals of opinion carried reviews of it: the *Bibliothèque universelle* in the Netherlands, the *Journal des sçavans* in France, and the *Acta eruditorum* in Germany. The review in the *Bibliothèque universelle*, almost certainly written by John Locke, merely summarized the work and indicated its place in the tradition of mathematical mechanics. The *Journal des sçavans* agreed that it presented "the most perfect mechanics that one can imagine," though it went on to object strenuously to the physical hypothesis it espoused – that is, the concept of attraction. Much the longest review appeared in the *Acta eruditorum*, an eighteen-page summary of the *Principia* expressed in a tone of warm admiration. Halley had also taken care to present copies to the leading philosophers of Europe. The *Principia* was not likely to go unnoticed.

Newton's book took Britain by storm. Almost at once it became the reigning orthodoxy among natural philosophers. On the Continent, its triumph was more protracted. Nevertheless, it refused to be ignored. Its impact can be gauged by the response of two towering figures, Christiaan Huygens and Gottfried Wilhelm Leibniz, both of whom received copies from Newton, and both of whom rejected its central concept. Huygens found the principle of attraction "absurd." For his part, Leibniz was astonished that Newton had not proceeded to find the cause of the law of gravity, by which he meant an aethereal vortex which would reduce attraction to a mechanical cause. Despite their carping, neither could conceal the impression the work had made. Huygens told his brother that he admired very much "the beautiful discoveries that I find in the work he sent me," and he made a point of meeting Newton when he visited England in 1689. Both men plied Fatio, their acquaintance, with questions about Newton and his work. Until Huygens's death in 1695, issues raised by the *Principia* – attractions, vortices, the shape of the earth, absolute motion, optics, mathematics – seasoned their correspondence. Titans such as Huygens and Leibniz did not become disciples of Newton. By dominating their correspondence, however, Newton demonstrated that the *Principia* had advanced him in one leap to the front rank of natural philosophers.

Other Continental philosophers indicated the same. Late in the 1690s, Dr. John Arbuthnot met the Marquis de l'Hôpital, a prominent French mathematician. L'Hôpital complained that none of the English could demonstrate to him the shape of a body that offered the least resistance to a fluid. When Arbuthnot showed him that Newton had done so in the *Principia* (as Conduitt recorded the story),

he cried out with admiration Good god what a fund of knowledge there is in that book? he then asked the D^r every particular about Sr I. even to the colour of his hair said does he eat & drink & sleep. is he like other men? & was surprized when the D^r told him he conversed chearfully with his friends assumed nothing & put himself upon a level with all mankind.

Meanwhile, events of a wholly different nature were thrusting on Newton another form of prominence which would affect the rest of his life, if not his enduring role in history, even more than the *Principia.* In 1687, as one of the first acts obligatory on a publisher, Halley presented a copy of the new book to James II, who had succeeded his brother Charles on the throne two years earlier, with a letter which dwelled especially on its treatment of the tides as a topic likely to interest an old naval commander. Probably James did not recognize the author's name. If he bothered to ask his advisers, he learned that the Lucasian professor of mathematics at Cambridge during the previous four months, while Halley shepherded the completed manuscript through the press, had placed himself irrevocably in the ranks of James's enemies.

The crisis, which had built up gradually for the university, finally arrived on 9 February 1687 in the form of a letter mandate to admit Alban Francis, a Benedictine monk, to the degree of Master of Arts without exercises and without oaths. The university had received many such mandates in the past and had readily conferred degrees on visiting Catholic dignitaries. Everyone understood that the case of Father Francis differed, however. Unlike the visitors, he intended to reside in Cambridge and, as a Master of Arts, to participate in the business of the university. No one could doubt that other fathers stood in the wings ready to descend on the university in Father Francis's wake to catholicize it. If the university were going to make a stand, it must be here.

Caught between the university and the court, John Peachell, the vice-chancellor, was beside himself. Almost frantic with anxiety, he wrote to his friend Samuel Pepys, an adviser to James, to explain his actions. "Worthy Sir, tis extraordinary distresse and affliction to me, after so much indeavour and affection to his Royall person, crown, and succession, I should at last, by the providence of God, in this my station, be thus exposed to his displeasure . . . ." Others we know of shared Peachell's concern. If we can believe Gilbert Burnet, whose account of these affairs is one of the primary sources about them, the anxiety so evident on the part of Peachell animated others; "all the great preferments of the church being in the king's disposal, those who did pretend to favour, were not apt to refuse his recommendation, lest that should be afterwards remembered to their prejudice."

The fears of Peachell help us to understand the events of 11 March, when, after a delay to consult and consider, the senate met. By 11 March, Newton was virtually free of the *Principia*. Halley had the manuscript of Book II, and Humphrey was copying Book III, which Newton would ship to London in three weeks. If he was free of the *Principia*, he was equally free of the concerns that immobilized others, for he had deliberately withdrawn from the pursuit of preferment nearly two decades before. We know nothing of what went on in the senate meeting on 11 March. We do know the result. The nonregent house (composed of the more senior Masters of Arts) chose Isaac Newton, a fellow Trinity hitherto known primarily for his aloofness from the university, as one of two representatives to convey to the vice-chancellor their voluntary advice that it would still be illegal and unsafe to admit Father Francis to the degree without the oath. Surely we must assume that Newton spoke out and articulated the common fears when prudential considerations left others mute. And in April, when a king furious at being thwarted summoned Peachell and representatives of the university to appear before the Court of Ecclesiastical Commission, the senate elected Newton (and Humphrey Babington) among the eight it designated to perform that duty.

Newton threw himself into the preparations for the hearing. His papers contain a number of documents, copied in Humphrey's hand, concerned with the defense. Newton himself told Conduitt that he alone had averted a compromise that would have surrendered the university's position. Before the delegation left for London, the chancellor of Ely drew up a paper in which its members agreed to admit Father Francis on condition that he not become a precedent for others,

to which they all seemed to agree but he disliking it arose from the table & took 2 or 3 turns & said to the Beadle who . . . was standing by the fire this is giving up the question, so it is said the beadle why do not you go & speak to it upon wch he returned to the table & told them his mind & desired the paper might be shewn to Council . . . .

In the end, he gave the delegation backbone enough to reject it. An identical account in Burnet, though without Newton's name, adds credence to the story.

On 21 April, while Halley was seeing the *Principia* through the press,

Newton with eight of his peers stood before the Ecclesiastical Commission led by the notorious Lord Jeffreys. In all, the delegation appeared before the commission four times, on 21 and 27 April and 7 and 12 May. Peachell was deprived of his academic offices, and the rest were treated to a lecture in which Jeffreys warned them that worse fates could fall upon them. But for all of that, Father Francis did not receive a degree – not least because Newton had refused to be frightened.

Once the revolution had ratified Newton's courage, he found himself one of the prominent figures in Cambridge. When the senate met on 15 January 1689 to elect two representatives of the university to the convention called to settle the revolution, Newton was one of three men put forward, and he was one of the two elected. From this time on, until he resigned his fellowship and chair in 1701, he was invariably one of the commissioners appointed by acts of Parliament to oversee the collection in Cambridge of aids voted to the government. Tax commissions were indexes to the leading citizens in a city or county, and Newton's appearance on them was a measure of his rising importance.

Newton also began to perceive himself in a new light which was incompatible with the isolation he had striven to maintain for twenty years. Humphrey Newton recalled that at his "seldom Entertainm^{ts}" his guests were primarily masters of colleges. We must assume that Humphrey referred to the period after the crisis of 1687. When David Loggan published his *Cantabrigia illustrata* in 1690, Newton appeared as the patron of the print of Great St. Mary's and in doing so placed himself in a circle of some eminence – including such men as the Duke of Lauderdale, the Earl of Westmoreland, Francis North, Baron Guilford, the bishops of Ely and of Lincoln, and Thomas Tenison, the future Archbishop of Canterbury. Nor was it the act of a retiring man when he arranged in London to have his portrait painted by the leading artist of the day, Sir Godfrey Kneller (Plate 1). The Kneller portrait is the earliest likeness of Newton we possess – an arresting presence, instinct with intelligence, caught when his capacities stood at their height. Without difficulty, we recognize the author of the *Principia*.

Newton set out for London almost as soon as he was elected to the convention. A note by Robert Morrice indicated that, together with Sir Robert Sawyer, the other representative of Cambridge, and Mr. Finch, perhaps the third candidate from Cambridge who was defeated in the election, he dined with no less a personage than William of Orange on 17 January. With the exception of six weeks in September and October, during an adjournment, he spent the whole of the following year in London.

There is no way to pretend that Newton played a leading role in the convention's deliberations. According to a story that rests solely on anecdotal authority, he spoke only once; feeling a draft, he asked an usher to close a window. It is not merely anecdotal that none of the surviving accounts of the Parliament contain any record of his participation in its debates. What evidence we have indicates that Newton stood squarely with the majority that declared James had forfeited the crown and that tendered it to William and Mary on 13 February.

Newton saw his principal role in the Parliament as forming a liaison with the university. During the early months of the session, he sent at least fourteen letters to the vice-chancellor, John Covel, with information about proceedings relevant to the university and advice on how the university should conduct itself. On one matter that certainly affected the university, the religious settlement, he did not say a word to Covel. We may assume that he was equally silent in Parliament; he had had a long lesson on the virtue of holding his tongue on that issue. Three hotly debated bills came before the Parliament: one to tolerate public worship by dissenters, one to repeal the Test Act of 1673, and one to comprehend many dissenters within the Church of England by broadening its definition. In the end, only the first passed into law. By its provisions, nearly all Protestant dissenters received the legal privilege to worship as they chose. Because the Test Act, which required that all public employees take the sacrament according to usage of the Anglican church, remained in effect, they did not receive civil equality. What must have concerned Newton most were the two exclusions from the privileges of toleration, Roman Catholics and "any person that shall deny in his Preaching or Writeing the Doctrine of the Blessed Trinity as it is declared in the aforesaid Articles of Religion [the Thirty-nine Articles of the Church of England]." The two exceptions were hardly equal. Protestant English people believed that Catholics threatened the sovereignty of the state. As the memory of James faded, their fears faded with it, and Catholics enjoyed toleration in fact if not in law. No one considered Arians a threat to the state. They were a threat rather to the moral foundations of society. Newton was well aware that the vast majority of his compatriots detested the views he held – more than detested, looked upon them with revulsion as an excretion that fouled the air breathed by decent persons. He had lived silently with that knowledge for fifteen years. The debate in Parliament, or the virtual lack of debate on a provision accepted without serious question, cannot have failed to bring it home to him once more.

Newton's heterodoxy allowed him easy concealment. Catholics aside, the laws had to do primarily with public worship. Newton did not worship in an Arian church. None such existed. As long as he was willing occasionally to take the sacrament of the Church of England, the law required of him nothing at which he need balk. Only on his deathbed did he venture finally to refuse the sacrament. Nevertheless, Newton had moved a considerable distance since 1674. In that year, he had prepared to vacate his fellowship rather than accept the mark of the beast in ordination. In his defense of the university in 1687 and in his service in Parliament, in both of which he pretended to orthodoxy, he demonstrated that his conscience had grown considerably less tender. Soon he began to court a position in London. It is manifest that he did not intend to let his religious convictions interfere. It is also true, as far as we know, that Newton did not seek reelection to Parliament in 1690. It has been assumed that distaste for the increasing factional strife in the final months of the session led him to withdraw. It may well be that the threat of discussion on matters he dared not discuss also played a role.

ๅ๏

The Parliamentary experience left no discernible mark on Newton. The year in London did. Freed from the constrictions of Cambridge society and buoyed by a new sense of confidence, he found new acquaintances under whose encouragement his accustomed reserve began to melt. One of them was Christiaan Huygens. Huygens's brother Constantyn had accompanied William of Orange on the expedition to England that ousted James. In June 1689, Christiaan came to visit him. On 12 June, within a week of his arrival in London, Huygens attended a meeting of the Royal Society and gave an account of his *Treatise of Light* and his *Discourse on the Cause of Gravity*, which he was about to publish together. Newton was present at the meeting. It strains credulity to believe that he was present by accident. The two met at least twice more, and before he left for home,

Huygens received two papers from Newton on motion through a resisting medium. At some point, they also discussed optics and colors. Huygens told Leibniz that Newton had communicated to him "some very beautiful experiments" on the subject – probably his experiments with thin films similar to the ones Huygens himself had performed less elegantly twenty years earlier. No continuing correspondence resulted from their meeting, however.

Continuing correspondence did result from another encounter. Although we do not know exactly when Newton met John Locke, it was probably during 1689 and probably at the house of the Earl of Pembroke. We know that they were already in correspondence before the autumn of 1690, the date on the first surviving letter, and a dated document implies that they knew each other well half a year earlier. The Newton of 1689 was a different man from the Newton of the 1670s. The completion and publication of the *Principia* and his own realization of its significance gave him new confidence. Nothing reveals the new Newton more clearly than his relation with Locke. Where he had shunned proferred correspondence with James Gregory, Huygens, and Leibniz in the 1670s, he not only seized the opportunity with Locke, but he did so with alacrity. The two shared many interests, all of Newton's commanding interests indeed with the exception of mathematics. Each recognized in the other an intellectual peer. Their letters from the early 1690s, exchanging views on the subjects Newton had pursued in isolation for nearly twenty years, marked a new departure in Newton's correspondence. Only his exchange with Boyle on chemical-alchemical matters, little of which survives, offered a precedent.

Religion provided what was easily the dominant theme of the correspondence and apparently of their conversation when they met. Locke later told his cousin Peter King that he knew few who were Newton's equal in knowledge of the Bible. On 14 November 1690, Newton sent Locke a treatise, in the form of two letters addressed to him, with the title *An historical account of two notable corruptions of Scripture, in a Letter to a Friend.* The two corruptions were the prime trinitarian passages in the Bible, 1 John 5:7 and 1 Timothy 3:16. Newton also composed a third letter about some twenty-six additional passages, all lending support to trinitarianism, that were corruptions too; we do not know if Locke ever received it. Although Newton presented the discourse as the mere disclosure of a pious fraud and not as a theological discourse, it is hard to believe that anyone in the late seventeenth century could have read it as anything but an attack on the trinity. Clearly Locke and Newton had got down to basics quickly and found they shared similar, unmentionable opinions. As far as we know, Newton had never dared to discuss his convictions with anyone before.

They shared as well a rationalistic approach to religion, which Newton had recently embodied in his "Origins of Gentile Theology." In his treatise on corruptions of scripture, Newton argued that 1 John 5:7 made sense without the disputed passage, but no sense with it.

If it be said that we are not to determin what's scripture & what not by our private judgments, I confess it in places not controverted: but in disputable places I love to take up wth what I can best understand. Tis the temper of the hot and superstitious part of mankind in matters of religion ever to be fond of mysteries, & for that reason to like best what they understand least. Such men may use the Apostle John as they please: but I have that honour for him as to beleive he wrote good sense, & therefore take that sense to be his wch is the best: especially since I am defended in it by so great authority.

It says much of Newton's confidence in 1690 that he sent such an Arian manifesto to Locke. It says even more that on the very morrow of the debate in Parliament he sent it with the explicit understanding that Locke was to forward it to the Netherlands to be translated into French and published – anonymously to be sure, but still published. Then as now, such matters had a way of not remaining secret. Locke accordingly mailed the treatise to Jean Le Clerc in Amsterdam, though without naming the author. A year later, Newton began to realize the immensity of the risk he was taking. Though the original understanding had been explicit enough, he now expressed surprise to learn that Locke had sent the manuscript forward, and he entreated him to stop the publication. He would pay any expenses that had been incurred. He was well advised. Le Clerc knew who the author was, and fifty years later, when his manuscript was found in the Remonstrants Library in Amsterdam where he deposited it, it was published under Newton's name. In 1692, such a publication would have led to Newton's ostracism from Cambridge and from society.

Early in 1692, another topic entered their correspondence. "I understand," Newton wrote in a postscript to a letter, "Mr Boyle communicated his process about ye red earth & ☿ [mercury] to you as well as to me &

before his death procured some of y^t earth for his friends." A continuing correspondence on alchemy, and on the exchange of alchemical information between Newton, Locke, and Boyle under pledges of secrecy, ensued. Much of the correspondence has been lost.

At about the same time he met Locke, Newton made another new acquaintance, Nicolas Fatio de Duillier (Plate 2). A brilliant Swiss mathematician, then only twenty-five years old, Fatio had come to England two years earlier after a stop in the Netherlands, where he had met Huygens. He carried an introduction from Henri Justel, a savant in Paris well known to the Royal Society, which promptly elected Fatio to membership. As a friend of Huygens, he attended the meeting on 12 June 1689 at which Huygens discoursed on light and gravity. There at least, if not before, he met Newton. The attraction between the two was instantaneous. Fatio was one of a party including Huygens, the Whig leader John Hampden, and Newton that set out from Hampton Court on 10 July to petition the king on Newton's behalf. On 10 October, a few days before his return for the second session of Parliament, Newton asked Fatio if there might be a chamber for him where Fatio stayed. "I intend to be in London ye next week & should be very glad to be in ye same lodgings wth you. I will bring my books & your letters wth me." Already the two were very close. By November, Fatio, who had arrived in England a Cartesian, had been converted to Newtonianism. Newton was (he wrote to his friend Jean-Robert Chouet) "*le plus honnête homme* I know and the ablest mathematician who has ever lived." He had discovered the true system of the world in a way that left no doubt for those who could comprehend it. The Cartesian system, which had revealed itself to Fatio as only "an empty imagination," was finished. If Newton had not already sent Humphrey home before he went to Parliament, his acquaintance with Fatio could have decided him to do so.

After Parliament was prorogued on 27 January 1690, Newton stayed on in London for another week. Near the end of February, Fatio wrote that he and John Hampden had planned to come to Cambridge to visit until Newton wrote that he was coming to London. He expected any day to receive the copy of the *Treatise of Light* that Huygens was sending to Newton. He would keep it until Newton told him to send it. "It beeing writ in French you may perhaps choose rather to read it here with me." The Exit and Redit Book of Trinity College indicates that Newton left on

10 March and returned again on 12 April. On 13 March, Fatio transcribed a revision of Proposition XXXVII, Book II, from Newton's copy of the *Principia*, and years later he mentioned a list of errata compiled by Newton which he had not had time to copy that March. We have every reason to think that Newton spent the month in London with Fatio, perhaps reading Huygens's *Treatise.*

About the beginning of June, Fatio went to the Netherlands for an extended stay of fifteen months, much of it with Huygens in The Hague. Newton wrote to Locke in October when he had not heard from him in half a year. When Fatio returned at the beginning of September 1691, he must have let Newton know at once, for, writing to Huygens on 8 September, he said he would see Newton soon, "since he is to come here in just a few days." Newton had scarcely returned home from London, where he had spent a month partly in the company of David Gregory and Edward Paget. The Exit and Redit Book indicates nevertheless that he was absent from the college from 12 to 19 September. He did not bother to get in touch with his other London friends. In October, Gregory wrote to tell him that Fatio had returned.

Although their correspondence during the following year, until September 1692, does not survive, Fatio's letters to Huygens indicate a steady exchange and at least one visit when Fatio saw some of Newton's mathematical papers. We know that Newton was in London much of January 1692; on 9 January, Pepys entertained him. Newton's letter to Fatio on 14 February 1693 mentioned a recent visit of Fatio in Cambridge. Huygens and Leibniz came to regard Fatio as their intermediary through whom they learned about Newton's opinions on mathematics, gravity, and light. Newton soon taught Fatio to share his other interests as well – heterodox theology, the prophecies, and alchemy – and they may have spent as much time on these matters as they did on mathematics and physics.

In May 1690, Newton received from Henry Starkey, whom he later described as his solicitor, a letter about a governmental appointment in London. Among other places, Starkey mentioned the positions of master, warden, and comptroller of the Mint, "very good places and they [the incumbents] make them as good as they please themselves . . . ." The crassness of the information does not appear to have offended Newton; a year later he wrote to Locke to ask him for a letter in regard to "the controulers place of yᵉ M." In fact, Locke with his political connections became Newton's principal agent in the search for an appointment. For his part, Newton continued to pursue a place in London with all the vigor he could muster. It is worth recalling that in the early 1690s Trinity lay in the grip of a financial crisis. In 1688, 1689, and 1690, it paid no dividend at all, and the two previous years, it paid only half dividends. A prudent man looked to his interests.

The growing realization that he had established himself as the leading intellectual of the land could not have discouraged Newton from seeking a position in the capital city. From every indication, he relished a new role of scientific consultant as much as he resented lesser intrusions on his time in earlier decades. In the summer of 1694, for example, the governing court of Christ's Hospital consulted him about a proposed revision of the curriculum of their mathematical school. Twelve years earlier, Newton's recommendation had played the major role in the selection of Edward Paget as master of the school. Now, perhaps in an effort to save a position he was about to lose because of neglect and dissolute habits, Paget proposed a revision of the school's curriculum. It was not Paget but the court which consulted Newton, however, and he devoted a great deal of time and trouble, composing several drafts, to his reply.

Moved to the forefront of English natural philosophy by the *Principia,* Newton began to be wooed by the younger generation, who sought his patronage. Whenever he was in London, he saw Edward Paget, though Paget was well on the way to forfeiting his position. On a visit in the summer of 1691, Newton met David Gregory, or Gregory finally met Newton after two efforts to establish a correspondence had failed. Gregory was the first to realize the potential advantage of Newton's favor, and from the first he courted him shamelessly. He flattered Newton extravagantly. "Farewell, noble sir," he concluded a letter, intended for publication to be sure, with a garbled line from Virgil, "and proceed as you do to advance philosophy 'beyond the paths of Sun and Sky'." (When the letter was not published, he did not blush to dust off the same line and send it to Huygens two years later.) Gregory's adulation was doubtless genuine. Even in the privacy of his personal memoranda, he referred to Newton only as "Mr. Newton" or, after 1705, "Sr Isaac Newton." Gregory also had a specific end in view, however. As a result of the resignation of Edward Bernard, the Savilian professorship of astronomy at Oxford stood empty. Gregory secured Newton's recommendation for the chair. As it happens, Edmond Halley also applied for the Savilian professorship. Despite his debt to Halley, a debt beyond payment, Newton not only refrained from supporting his application but also did support Gregory. Gregory got the chair. The lesson was not lost on other aspiring young men. Nor was it lost on Gregory, who continued to court Newton assiduously.

Not long before he left Cambridge, another hopeful young man, William Whiston, took care to make his acquaintance. By his own account, Whiston heard one or two of Newton's lectures on the *Principia* while he was an undergraduate and failed to understand them. He set himself to master Newtonian philosophy in the early 1690s, and in 1694 he submitted the manuscript of his *New Theory of the Earth* to Newton's inspection. According to Whiston, it won approval. For the time being, nothing more that we know of came of their relation, though Whiston may have discussed theology with Newton, who was finding that others were prepared to entertain his doubts about trinitarian orthodoxy. Not long thereafter, at any rate, Whiston became the articulate spokesman for views virtually identical to Newton's. In 1701, when Newton finally resigned the Lucasian chair, he secured Whiston's nomination to succeed him. Probably with that end in view, he had appointed him his deputy a few months earlier.

By 1701, Newton had filled two of the three university chairs devoted to science and mathematics with his disciples. Shortly, he would place Halley in the other Savilian chair at Oxford and help to establish another disciple in the new Plumian chair in Cambridge.

₴

As the new dimensions of his existence gradually unfolded, Newton did not forsake his intellectual pursuits. Quite to the contrary, the early years of the 1690s, his final years in Cambridge, were a period of intense, almost manic intellectual activity. Swept along by the *Principia*'s success, Newton attempted to pick up the scattered loose threads of earlier enquiries and to weave them into a coherent whole worthy of his great completed work. The effort was his last major intellectual endeavor.

Interestingly, theology was not prominent among his pursuits during this period. If we judge by the surviving papers, the *Principia* interrupted the study which, with alchemy, had dominated his attention during the previous fifteen years, and he did not return to it seriously for another two decades. It is true that external factors could elicit theological opinions from him, and there is good reason to think that he discussed the subject with a number of trusted intimates, such as Fatio, Halley, and Whiston, who were later reputed or known to be Arians. What Newton had to say in such discussions stemmed from his earlier conversion to heresy, not from current study.

In 1693, Newton engaged in a quasi-theological correspondence with Richard Bentley, an aspiring young cleric of formidable intellect. Bentley was named to deliver the first set of lectures in defense of religion established by the will of Robert Boyle. Late in 1692, as he prepared the manuscript of his lectures (which had drawn heavily on Newton) for publication, he applied to Newton for help with several points. In all, Newton sent four letters on the subject to Bentley.

When I wrote my treatise about our Systeme [he began the first one] I had an eye upon such Principles as might work wth considering men for the beleife of a Deity & nothing can rejoyce me more then to find it usefull for that purpose. But if I have done ye publick any service this way 'tis due to nothing but industry & a patient thought.

He went on to summarize the reasons that convinced him that the universe as we know it could not have resulted from mechanical necessity alone but required the intelligence of a Creator. "There is yet another argument for a Deity wch I take to be a very strong one," he concluded enigmatically, "but till ye principles on wch tis grounded be better received I think it more advisable to let it sleep." As far as I know, Newton never explained this reference. He probably had in mind the argument from the providential course of history as foretold in the prophecies.

During the early 1690s, Newton also gave thought to putting his mathematical achievement into publishable form. Leibniz had begun to publish his differential calculus in the autumn of 1684, not mentioning Newton in any of the papers he put out. Leibniz can be excused. Newton had nothing on mathematics in print, and a reference to him would not have meant anything to most European mathematicians. Nevertheless, in view of the correspondence of 1676, one would not care to cite Leibniz's silence as a lesson in generosity. Nor would one call it judicious – as

Leibniz must later have reminded himself ruefully many times. Although no previous allusions to Leibniz's publications appeared in Newton's papers, he seems to have nursed a growing sense of grievance. For as soon as he began to write a letter about the binomial expansion that Gregory requested in the fall of 1691, he forgot Gregory, returned forthwith to the correspondence of 1676, and began to compose a defense of his priority vis-à-vis Leibniz.

The letter rapidly converted itself into a full-scale exposition of Newton's fluxional method, "De quadratura curvarum" ("On the Quadrature of Curves"), which began with a recitation of his exchange with Leibniz in 1676 and went on to present an extended set of problems that the fluxional method could solve, problems like those to which Leibniz was addressing his calculus. As though in conscious competition, it developed for the first time on Newton's part a systematic notation as an alternative to Leibniz's. It was "De quadratura" that adopted the familiar dot notation for fluxions, and it experimented with Q (for *quadratura*) as a substitute for Leibniz's ∫ (for *summa*) as a symbol for the operation of squaring.

By the end of 1691, Newton's circle of young friends in London had become aware of his treatise. However, Newton's interest waned as rapidly as it had waxed. By March, Fatio reported to Huygens that Newton's enthusiasm had passed and that he had begun to think he might better avoid the embarrassments that its publication was bound to cause. "We shall assuredly lose a great deal if this Treatise does not appear," he added. "It is certain that until now nothing in abstract Geometry as beautiful as this writing has ever appeared . . . ." Eventually Newton published a truncated version of "De quadratura" as an appendix to his *Opticks*.

His friends in London had also picked up the priority issue. On 18 December 1691, Fatio stated the issue to Huygens bluntly.

It seems to me from everything that I have been able to see so far, among which I include papers written many years ago, that Mr. Newton is beyond question the first Author of the differential calculus and that he knew it as well or better than Mr. Leibniz yet knows it before the latter had even the idea of it, which idea itself came to him, it seems, only on the occasion of what Mr. Newton wrote to him on the subject. (Please Sir look at page 235 [Lemma 11, Book II] of Mr. Newton's book). Furthermore, I cannot be sufficiently surprised that Mr. Leibniz indicates nothing about this in the Leipsig Acta [his papers that published the differential calculus].

By February, Fatio was more explicit.

The letters that Mr. Newton wrote to Mr. Leibniz 15 or 16 years ago speak much more positively than the place that I cited to you from the Principles which nevertheless is clear enough especially when the letters explicate it. I have no doubt that they would do some injury to Mr. Leibniz if they were printed, since it was only a considerable time after them that he gave the Rules of his Differential Calculus to the Public, and that without rendering to Mr. Newton the justice he owed him. And the way in which he presented it is so far removed from what Mr. Newton has on the subject that in comparing these things I cannot prevent myself from feeling very strongly that their difference is like that of a perfected original and a botched and very imperfect copy. It is true Sir as you have guessed that Mr. Newton has everything that Mr. Leibniz seemed to have and everything that I myself had and that Leibniz did not have. But he has also gone infinitely farther than we have, both in regard to quadratures, and in regard to the property of the curve when one must find it from the property of the tangent.

Although Newton decided to avoid the unpleasantries that he saw his treatise would entail, the priority issue did not wholly disappear. In the summer, John Wallis opened the pages of his forthcoming *Opera* to whatever Newton might wish to insert. Before long, Wallis began to pester Newton without interruption about Leibniz. He reported then that he had heard from Holland that "your Notions (of *Fluxions*) pass there with great applause, by the name of *Leibnitz's Calculus Differentialis.*" Newton had not forgotten him. The précis of "De quadratura" that he sent to Wallis addressed itself primarily to the German mathematician. It began with the same reference to René de Sluse in the *Epistola posterior*. It translated both of the anagrams in that letter. And in presenting briefly Newton's solutions to fluxional equations in connection with the second anagram, it pretended that already in 1676 Newton had developed methods that did not in fact appear in his papers before the 1690s. Newton also asked Wallis to insert his series for the circle, as given in the letter of 1676, beside Leibniz's series for the circle, which was also to appear in the volume. Manifestly Newton still had the matter in mind.

So did Leibniz. Huygens had passed on some of Fatio's assessments of Newton's achievements, though not his comments on Newton's priority, and mathematicians all over Europe heard of his plans to publish. When the plans boiled down to an exposition of his method in Wallis's *Opera,* they waited with anticipation for that, Leibniz among them and for his own reasons. In March 1693, he wrote to Newton – a gracious letter seeking, as he had sought nearly twenty years before, to initiate a philosophical correspondence, but also a nervous letter.

How great I think the debt owed to you, by our knowledge of mathematics and of all nature, I have acknowledged even in public when occasion offered [he began with something less than complete candor]. You had given an astonishing development to geometry by your series; but when you published your work, the *Principia,* you showed that even what is not subject to the received analysis is an open book to you.

He went on to refer to his own work in mathematics. "But to put the last touches I am still looking for something big from you . . . ." From mathematics, he proceeded to the *Principia* and to what he had heard from Huygens about Newton's optics. In conclusion, he acknowledged his long silence, though he might well have complained instead of Newton's, and excused himself by saying he had not wanted to burden Newton with letters.

Personal problems postponed Newton's answer until October, and he opened with an apology for the delay. "For although I do my best to avoid philosophical and mathematical correspondences, I was nevertheless afraid that our friendship might be diminished by silence . . . ." He feared it all the more, he continued, because Wallis had just inserted some new points from their earlier correspondence in his forthcoming work. At Wallis's request, he was revealing the method concealed before in the anagram, the translation of which he included in the letter – a limited favor, as he had already published it in the *Principia.* "I hope indeed that I have written nothing to displease you, and if there is anything that you think deserves censure, please let me know of it by letter, since I value friends more highly than mathematical discoveries." Although he made a brief response to Leibniz's other comments, the letter was the thinnest answer possible, and Leibniz made no attempt to continue correspondence.

The veiled threat in the coming publication could not be missed. In June 1694, Leibniz had still not seen Wallis's volume, and he wrote impatiently to Huygens to send it as soon as possible. When he finally received it in September, he expressed his disappointment that it contained so little on the inverse problem of tangents, but the disappointment sounded more like relief. The two methods were similar, he remarked, but his own was clearer. All Newton presented on the inverse problem of tangents was a means of expressing a given ordinate by an infinite series, which he had understood at the time – that is, in 1676. In a word, Leibniz's attention also focused on the priority issue, and the publication did not seem to undercut him as he had feared. Johann Bernoulli read it in the same light. Writing to Leibniz, he questioned whether Newton had not in fact plundered Leibniz's publications to fashion the method which he only now presented. If the potential dispute which flickered and nearly burst into flame seemed to die down, it was far from dead.

Nor was it in England. Among the memoranda that Gregory made of his conversations with Newton in the summer of 1694, he recorded an ominous question: "Whence the differential calculus of Leibniz." And in the fall, he sketched out a treatise on the calculus that he intended to write in which he would show that Leibniz's calculus reduced to Newton's, which alone had been fully demonstrated. Wallis also was unwilling to let the issue die. Disappointed with what he had been able to publish in the first two volumes of his *Opera,* he began to press Newton in 1695 for permission to publish the two *Epistolae* in full. In May, he sent Newton the copies of them that he had had since the 1670s and asked Newton to correct them for publication; when Newton did not reply, he wrote again in July. This time Newton complied. He also thanked Wallis "for your kind concern of right being done me by publishing them." To indicate that he was also concerned, he quoted a passage from Collins's letter of 18 June 1673 to insert as a note to the reference to Sluse. If the *Epistola posterior* were not enough, the note should do the job. The letter it quoted, dated before the publication of Sluse's paper, asserted Newton's possession of the method at that time. "I would willingly see Leibnitz's answer," Wallis remarked to Halley.

Along with the *Principia* and mathematics, Newton returned to his optics during the late 1680s and early 1690s after an interval of nearly two decades. He drew his work in optics together into a volume but decided then not to publish it. Nor did he omit alchemy as he drew his philosophic legacy together. On the contrary, a good half of his extensive alchemical papers, which were interrupted for two or three years by the *Principia,* came from the period immediately following it. If we can judge by the quantity of manuscripts, Newton devoted more of his time to alchemy in the early 1690s than he did to everything else put together.

From his alchemy, Newton drew a paper, "De natura acidorum" ("On the Nature of Acids"), which he allowed Archibald Pitcairne, Gregory's friend, to have when he visited Cambridge early in March 1692. Pitcairne also took extensive notes on their conversations which supplement the short essay. The particular value of "De natura acidorum" lies in the glimpses it affords of Newton translating the activities of alchemical principles into the vocabulary of forces. "The particles of acids," he asserted, ". . . are endowed with a great attractive force and in this force their activity consists by which they dissolve bodies and affect and stimulate the organs of the senses." During the early 1690s, Newton also continued to compose alchemical essays. It was during this period that he set down the most important alchemical essay he ever wrote, "Praxis," a treatise which employed all the imagery of the alchemical tradition and brought in most of the substances of his earlier experimentation – the net, the oak, sophic sal ammoniac, the doves of Diana, the star regulus of Mars. "Praxis" reached its climax with the ultimate alchemical claim "Thus you may multiply to infinity."

From a reference to Fatio's letter of May 1693, which he added to a draft, Newton appears to have composed "Praxis" in the spring and summer of 1693, a time of great emotional stress. We should probably read its extravagant claim in the light of that stress. We should recall also that disillusionment apparently as full as his earlier expectations followed not long after and may have heightened the stress in 1693. In 1681, Newton had crossed out his two climactic exclamations of success. He did not literally cancel "Praxis," but he did so implicitly by abandoning it. It is tempting to connect the disillusionment with the fateful year 1693, though evidence such as dated experiments in 1695 and 1696 tells us that it could not have been complete in that year, if indeed it began then at all. That there was a disillusionment, however, and that it came not long after 1693 cannot well be ignored. I have found only four alchemical notes, all fragmentary, that can be dated with assurance to the period after Newton's move to London. Whereas Newton extended all of his other major interests into the London residence, he devoted no further significant time to alchemy. He did continue to acquire some alchemical books; he did not, as far as surviving documents inform us, take notes from them. Conduitt later recorded one nostalgic comment, that "he would if he was younger have another touch at metals." It suggested as little sustained interest as the unread books. Nearly thirty years of intense devotion to the Art left their permanent mark on Newton's intellect. Nevertheless, they did come to an end.

≈

Meanwhile, in 1693, more than Newton's alchemy reached a climax. His relations with Fatio did also. Fatio visited Newton during the autumn of 1692. Presumably not long after he left, Newton received a letter from him written on 17 November.

I have Sir allmost no hopes of seeing you again. With coming from Cambridge I got a grievous cold, which is fallen upon my lungs. Yesterday I had such a sudden sense as might probably have been caused upon my midriff/diaphragm by a breaking of an ulcer, or vomica, in the undermost part of the left lobe of my lungs. . . . My pulse was good this morning; It is now (at 6. afternoon) feaverish and hath been so most part of the day. I thank God my soul is extreamly quiet, in which you have had the chief hand. . . . Were I in a lesser feaver I should tell You Sir many things. If I am to depart this life I could wish my eldest brother, a man of an extraordinary integrity, could succeed me in Your friendship.

A reply frantic with concern arrived at once. ←

S^r

I . . . last night received your letter wth wch how much I was affected I cannot express. Pray procure ye advice & assistance of Physitians before it be too late & if you want any money I will supply you. I rely upon ye character you give of your elder brother & if I find y^t my acquaintance may be to his advantage I intend he shall have it . . . S^r wth my prayers for your recovery I rest

> Your most affectionate
> and faithfull friend to serve you
> Is. Newton

Fatio had dramatized a cold excessively. He was recovering by the time Newton's letter arrived, and in fact he lived for another sixty-one years.

Nevertheless, the cold did linger. In January, a Swiss theologian then in England, Jean Alphonse Turretin, brought Newton word of Fatio's continuing illness, and by the same hand Newton sent a radical proposal back to Fatio.

I feare ye London air conduces to your indisposition & therefore wish you would remove hither so soon as ye weather will give you leave to take a journey. For I beleive this air will agree wth you better. Mr Turretine tells me you are considering whether you should return this year into your own country. Whatever your resolutions may prove yet I see not how you can stirr w^{th}out health & therefore to promote your recovery & save charges till you can recover I [am] very desirous you should return hither. When you are well you will then know better what measures to take about returning home or staying here.

Fatio confirmed his intention to return to Switzerland. The recent death of his mother made the trip more necessary. With the estate she had left him, he added, he could live for some years in England, chiefly in Cambridge, "and if You wish I should go there and have for that some other reasons than what barely relateth to my health and to the saving of charges I am ready to do so; But I could wish in that case You would be plain in your next letter."

The correspondence continued on through the winter and into the spring, focusing on Fatio's health and finances and warily circling the question of his possible move to Cambridge. As to the saving of charges, Newton mentioned his plan "to make you such an allowance as might make your subsistence here easy to you." At the time he offered the allowance, Newton had not received a full dividend from Trinity for seven years. Convinced that Fatio was destitute, he forced an extravagant sum upon him for a couple of books and some medicine that Fatio had left in Cambridge, and in May he offered him more money. Beyond the issue of money he could not be plain, however. Perhaps Fatio tried. "I could wish Sir," he wrote in April, "to live all my life, or the greatest part of it, with you, if it was possible, and shall allways be glad of any such methods to bring that to pass as shall not be chargeable to You and a burthen to Your estate or family." He went on to add that Locke had just been there pressing a proposal that the two of them settle with him at the Masham estate in Essex. "Yet I think he means well & would have me to go there only that You may be the sooner inclinable to come."

In May, Fatio's correspondence turned to alchemy, a subject to which Newton had introduced him. He had met a man who knew a process by which gold amalgamated with mercury vegetates and grows. With his usual sense of drama, he told Newton to burn the letter as soon as he had read it. Less impressed, Newton kept the letter, though he did insert a reference to the process in his "Praxis." Two weeks later, Fatio's new friendship had flourished as well as the matter in the glass. His friend also made a medicine from his mercury; it had finally cured him. The friend now proposed that they become partners in its production. Fatio would need to spend a couple of years picking up a medical degree. With the degree in hand together with the medicine, which was very cheap, he could cure thousands of people for free in order to make the medicine known. It was good for consumption and smallpox, and it freed the body of atrabile (black bile), which was known to cause nine diseases out of ten. He could make a fortune. There was a hitch, however. He would need between £100 and £150 per year for at least four years; and in a hesitant, backhanded way, he let Newton know that now was the time to help him financially. He urgently requested also that Newton come to London to advise him.

On 30 May, Newton signed out of Trinity for a week. Undoubtedly he went to London, as concerned with Fatio's new friend as he was with his finances. Apparently he left Trinity for another week late in June, no doubt to go to London again. Two other pieces of information complete our meager budget of knowledge about this critical summer. On 30 May, he started a letter to Otto Mencke, the editor of the *Acta eruditorum,* which he set aside and did not complete for six months. He experimented in his laboratory in June. Beyond that, silence covered nearly four months. He broke silence with a letter to Samuel Pepys on 13 September.

Sir,

Some time after Mr Millington had delivered your message, he pressed me to see you the next time I went to London. I was averse; but upon his pressing consented, before I considered what I did, for I am extremely troubled at the embroilment I am in, and have neither ate nor slept well this twelve month, nor have my former consistency of mind. I never designed to get anything by your interest, nor by King James's favour, but am now sensible that I must withdraw from your acquaintance, and see neither you nor the rest of my friends any more, if I may but leave them quietly. I beg your pardon for saying I would see you again, and rest your most humble and most obedient servant,

Is. Newton

Three days later, now at an inn in London, he wrote to John Locke.

S^r

Being of opinion that you endeavoured to embroil me wth woemen & by other means I was so much affected with it as that when one told me you were sickly & would not live I answered twere better if you were dead. I desire you to forgive me this uncharitableness. For I am now satisfied that what you have done is just & I beg your pardon for my having hard thoughts of you for it & for representing that you struck at ye root of morality in a principle you laid down in your book of Ideas & designed to pursue in another book & that I took you for a Hobbist. I beg your pardon also for saying or thinking that there was a designe to sell me an office, or to embroile me. I am

your most humble & most
unfortunate Servant
Is. Newton

Eighteen months earlier, the search for a position in London had given rise to paranoid passages in Newton's correspondence with Locke. "Being fully convinced that M^r [Charles] Mountague upon an old grudge wch I thought had been worn out, is false to me," he had written in January 1692, "I have done wth him & intend to sit still unless my Ld Monmouth be still me friend." Three weeks later, he had expressed his pleasure at Lord Monmouth's continued friendship and excused himself with exaggerated concern for an imagined infelicity the last time he saw Monmouth – a letter expressing groveling subservience which is embarrassing to read as a product of Newton's pen. In withdrawing his essay on the corruption of Scriptures, which he had confidently given to Locke for publication a year earlier, the same two letters had marked the end of the manic euphoria that had borne Newton forward since the twin triumphs of the *Principia* and the revolution. Now, in the autumn of 1693, what Frank Manuel has aptly named Newton's black year, he plumbed the depths of the ensuing depression.

The following May, a Scotsman named Colm told Huygens that Newton had had an attack of frenzy that had lasted eighteen months. It was thought that, in addition to excessive study, a fire that had destroyed his laboratory and some of his papers had contributed to troubling his mind. Friends had confined him until he had recovered enough to recognize his *Principia* again, but Huygens assumed he was lost to science. Huygens's story of the fire may appear to confirm the passage quoted earlier from the diary of Abraham de la Pryme, although the dates impose a major problem of reconciliation because Pryme recorded his account in February 1692. In any case, Huygens repeated the story to Leibniz, and it spread rapidly through the whole European community of natural philosophers. By the summer of 1695, John Wallis received from Johann Sturm in Germany an account, which he bluntly denied, that Newton's house and books had all burned and that Newton himself was "so disturbed in mind thereupon, as to be reduced to very ill circumstances."

The story of the fire carries doubtful authority. The letters to Pepys and Locke were undoubtedly authentic, however, and it is impossible to pretend that Newton did not go through a period of mental derangement – though not necessarily one like that described to Huygens and not one that lasted, in its most acute stage, for eighteen months. One cannot sufficiently admire Pepys and Locke. Confronted without warning by such letters, neither gave thought to taking offense. Rather both assumed at once that Newton was ill and acted accordingly. Through his nephew, who was a student in the university, Pepys enquired discreetly of John Millington, the fellow of Magdalene College mentioned in Newton's letter, and eventually wrote directly to Millington. Pepys questioned Newton's sanity. After the visit from Pepys's nephew, Millington, who assured Pepys that he had not delivered any message at all to Newton, much less the one Newton alleged, had tried to call on Newton but found him out. He met him finally on 28 September in Huntingdon,

> where, upon his own accord, and before I had time to ask him any question, he told me that he had writ to you a very odd letter, at which he was much concerned; added, that it was in a distemper that much seized his head, and that kept him awake for above five nights together, which upon occasion he desired I would represent to you, and beg your pardon, he being very much ashamed he should be so rude to a person from whom he hath so great an honour. He is now very well, and, though I fear he is under some small degree of melancholy, yet I think there is no reason to suspect it hath at all touched his understanding . . . .

Pepys waited two months until Neale's lottery offered an occasion to write. After a brief and dignified reference to Newton's letter, he assured him of his readiness to serve him and proceeded forthwith to test his understanding for himself by putting the question of chance. The response would not have left him in doubt that Millington's assessment was correct.

Locke, who had no nephew to press into service, waited for two weeks before he replied early in October.

> I have ben ever since I first knew you so intirely & sincerely your friend & thought you so much mine y^t I could not have beleived what you tell me of your self had I had it from anybody else. And though I cannot but be mightily troubled that you should have had so many wrong & unjust thoughts of me yet next to the returne of

good offices such as from a sincere good will I have ever done you I receive your acknowledgm^t of the contrary as ye kindest thing you could have done me since it gives me hopes I have not lost a freind I soe much valued.

Newton answered with much the same explanation that he gave to Millington.

The last winter by sleeping too often by my fire I got an ill habit of sleeping & a distemper wch this summer has been epidemical put me further out of order, so that when I wrote to you I had not slept an hour a night for a fortnight together & for 5 nights together not a wink. I remember I wrote to you but what I said of your book remember not.

Later in the fall, when he answered letters from Leibniz and Mencke that he had received six months earlier, Newton told them both that he had mislaid their letters and had not been able to find them.

Over the years a variety of explanations have been offered for Newton's breakdown. A variant of an old theory, which blamed it on exhaustion from the strain of composing the *Principia*, appears plausible to me. As I have argued, the early 1690s were a period of intense intellectual activity for Newton as he strove to weave the various strands of his disparate endeavors into a coherent fabric. Intellectual excitement always stretched him to the very limit and, on occasion, beyond it. His breakdown in 1693 was not altogether different from his behavior in 1677–8. Neither episode was incommensurable with what we know of an isolated student, alienated from his peers, already in his undergraduate years. If 1693 was a climax, it was a climax long in preparation. In the early 1690s, there was the added stress of a manifest sense of humiliation in soliciting an appointment in London. Add also the growing doubts that had begun to assail him. By 1693, he had withdrawn his theological publication, scuttled plans to publish works in mathematics and optics, and begun to hesitate on a second edition of the *Principia*. The apparent climax and disillusionment in alchemy in the summer of 1693 only confirmed his doubts. There may also have been a fire – another fire, as it appears to me – which could well have distracted him when he was already in a state of acute tension. Charred papers from the early 1690s survive, though it is difficult to fit them satisfactorily with the other dated events.

It is also impossible to leave Fatio out of the account. Newton was not the only one who was in turmoil. Fatio was also going through a period of acute personal and religious tension. The sense of approaching crisis became almost palpable in their correspondence in early 1693. It is unlikely that we will ever learn what passed between them in London. Their relation ended abruptly, however, never to be resumed. The primary focus of Newton's attention for four years, from the time they met in 1689, Fatio simply dropped out of Newton's life. The rupture had a shattering effect on both men. Newton rebounded from his breakdown, but Fatio effectively disappeared from the philosophic scene. For a number of years he hovered on the edge of intellectual circles without really being in them. In 1699, he reappeared on the stage briefly with a mathematical tract which, in a reference to Leibniz possibly intended to regain Newton's favor, fanned back into flame the languishing priority dispute over the calculus. In the early eighteenth century he plunged into the fanatical Camisard prophets from France and disappeared for good from the community of natural philosophers among whom his star had seemed destined to blaze. Beyond his role in rekindling the calculus controversy, he played no further part in Newton's life.

When Newton's breakdown first became known to historians early in the nineteenth century, Jean Baptiste Biot interpreted it as the turning point in his life that brought his scientific activity to an end and inaugurated his theological studies. In the form that Biot presented it, the interpretation cannot stand. When David Gregory visited Newton in May 1694, he could hardly write fast enough to take down notes of projects on which Newton was at work or at least pretended plausibly to be. Newton's mathematical papers do not indicate a break in 1693; significant ones certainly dated from the following years. In 1694, he took up anew one of the most difficult problems in the *Principia*, the lunar theory. Manifestly, Newton was still capable of clear scientific thought on the most perplexed question. As for theology, we have seen that his greatest sustained attention to it belonged to the fifteen years that preceded the *Principia*. Nevertheless, a revised version of Biot's thesis does appear correct. The year 1693 witnessed the climax of the intense intellectual effort that followed the *Principia;* and if Newton had by no means lost his mental coherence after 1693, it remains true that he did not again inaugurate any new investigation of importance. He was no longer a young man. The crisis of

1693 terminated his creative activity. In theology as well as in natural philosophy and mathematics, he devoted the remaining thirty-four years of his life to reworking the results of earlier endeavors – insofar as he did not take refuge in administrative activity to absorb his time.

≀≀

The possible exception to the assertion above was Newton's work on lunar theory, which dominated his attention for a year beginning in the summer of 1694. His effort takes on special significance against the background of 1693. The moon presented the most complex problem in his *Principia*. The first edition had defined an attack on the three-body problem and had made a start at bringing the manifold perturbations of the lunar orbit, hitherto known only empirically insofar as they were known at all, under quantitative treatment within the theory of gravitation. Neither he nor Halley had been satisfied. Now he tried to make his treatment more precise so that, as he told Flamsteed and Gregory, the discrepancy between theory and observation would not exceed two or three minutes of arc. It was excruciating work. The three-body problem does not admit a general analytic solution, and he had to deal with a set of corrections which could be distinguished from one another and defined quantitatively only with difficulty. He later told Machin that "his head never ached but with his studies on the moon."

In order to complete the task he needed observations that only Flamsteed could furnish. Whatever else he might be, Flamsteed was not a remedy for a headache. Nor was Newton medicine for Flamsteed's ailments. For most of a year two difficult men managed to restrain their impatience and deal with each other. Newton harassed Flamsteed for the observations he needed; Flamsteed, intent on his own program of observation, received the requests, which he did his best to fulfill, as so many interruptions. As the lunar problem itself began to baffle Newton, he projected his frustrations and headaches onto Flamsteed and his unwillingness to supply the needed observations. A furious letter of July 1695, in which Newton tried to shift his own failure onto Flamsteed, marked the end of his effort.

The failure seemed to confirm the breakdown of 1693. Late in 1695, Newton entered into a vigorous correspondence with Halley on comets. For eight years there had been little commerce between the two men.

Reading the lesson of Gregory's triumph at his expense, Halley appears to have decided that he must cultivate Newton more explicitly. From Newton's point of view, the correspondence did not break significant new ground and did not compensate for the lunar debacle.

If he had not realized the high hopes of the early 1690s, Newton's achievement during these years had nevertheless been considerable. He had put the *Opticks,* the second of the twin pillars of his reputation in science, into the form that he later published. He had completed the two mathematical papers that he himself later published, which gave substance to his reputation as a mathematician. Nevertheless, he had not achieved the great synthesis at which he had grasped. The turning point of the *Principia* had come too late. Newton was now well over fifty. He knew that his powers had begun to fade. If it was too late to crown the triumph of the *Principia,* his academic sanctuary ceased to have meaning. He had a second string to his bow, however. He had not yet received his reward for the triumph of the Glorious Revolution.

Against this background, we can evaluate his decision of 1696, so incomprehensible to the twentieth-century academic mind, to leave Cambridge for a relatively minor bureaucratic post in London. The notion was not new to him. He had pursued an appointment fruitlessly in the early 1690s. Now, with his friend Charles Montague advancing in power, it became a possibility once more. What attraction would hold him in Cambridge? Certainly not the intellectual community. He had never found such there and had consciously held himself aloof from his peers. London had supplied his first real experience of an intellectual community; and if a desire for such played any role in his decision, it must have tilted the scale decisively for London. Cambridge's advantage for him had always been the uninterrupted leisure it provided to pursue his studies. As he felt his creative energy ebbing, that advantage evaporated. Indeed, his failures in the 1690s may well have driven him to escape from unproductive leisure into concrete activity.

What is equally relevant is that Cambridge supplied a rationale. Newton had not separated himself so far from the university that he remained untouched by its ethos. To the Restoration don, the institution existed, not to be served, but to be exploited for personal advantage. In the early 1690s, Newton's peers in Trinity were at last approaching the ultimate status of senior fellows. George Modd, Patrick Cock, Nicholas Spencer, and William Mayor had never among them tutored a student, taken an advanced degree, or produced a line of scholarship. Nevertheless the system of seniority had carried them always abreast of Newton, and now they were beginning to reap their rewards. All now enjoyed lucrative college livings in the environs of Cambridge to supplement their fellowships. The college nominated them in their turns for university offices such as taxor and scrutator, which provided additional income. They had used the university to build their fortunes and had given nothing in return. Newton had stood resolutely aside from the scramble for ecclesiastical preferment, but he indicated that he understood the reigning mores when he silently converted his chair to a sinecure following 1687. He would now demonstrate to those who in amusement repeated anecdotes about the strange fellow who lived by the gate that ecclesiastical advancement was not the only – or the richest – reward one could extract.

Stories about an appointment circulated in London. In November 1695, Wallis heard that Newton was to be master of the Mint. Newton denied another report roundly to Halley on 14 March 1696.

And if the rumour of preferment for me in the Mint should hereafter upon the death of Mr Hoar or any other occasion be revived, I pray that you would endeavour to obviate it by acquainting your friends that I neither put in for any place in the Mint nor would meddle with Mr Hoar's place were it offered me.

Fortunately, Halley did not have to waste much effort denying the stories. At the very time Newton wrote, Montague was completing the arrangements to appoint him, not to Hoare's post of comptroller, to be sure, but to the better one of warden. Montague dated his letter confirming the appointment 19 March. Newton accepted without pausing even to reflect. After a residence in Trinity of thirty-five years, he contrived to depart, bag and baggage, in less than a month, part of which he spent in London. Although he continued to hold both fellowship and chair, and to enjoy their incomes, for another five years, he returned only once for half a week to visit. As far as we know, he wrote not a single letter back to any acquaintance made during his stay.

10

The Mint

CRISES RACKED THE INSTITUTION to which Newton moved in the spring of 1696. Indeed, the Mint was an institution within an institution within an institution, all three of which faced crises. The recoinage engaged every pinch of energy at the Mint. The Treasury, of which the Mint was a relatively minor department, devoted equal energy to devising temporary expedients and new machinery to cope with overwhelming financial needs caused by war with France. The English state and the revolutionary settlement it embodied balanced precariously on the outcome of the Treasury's efforts. In 1696, it was not clear that the financial demands of the war would be met. If they were not, if national bankruptcy ensued, the revolutionary settlement would undoubtedly collapse before a second Stuart restoration. In the larger crises of the government and its finances Newton was not involved beyond his concern as an Englishman committed to the revolution.

The narrower monetary crisis, which bedeviled the financial crisis by reaching a climax when it could least be tolerated, occupied him almost completely for more than two years. As the debasement of silver coinage reached disastrous proportions, the government under the leadership of Newton's friend Charles Montague, then Chancellor of the Exchequer, began to consider a recoinage as the only effective remedy.

In 1695, the government sought what advice it could find. In the absence of a body of recognized experts on such matters, the Regency Council resolved to consult a number of leading intellectuals and London financiers, "Mr. Locke, Mr. D'Avenant, Sir Christopher Wren, Dr. Wallis, Dr. Newton, Mr. Heathcote, Sir Josiah Child, and Mr. Asgill, a lawyer." Along with most of the others, Newton replied in the autumn of

1695 with a short essay "concerning the Amendm^t of English Coyns." A general consensus among the respondents, in which he shared, accepted the need to recoin. By December, the decision was made; on 21 January 1696, the definitive recoinage act passed through Parliament. The first melting of old coins at the Exchequer commenced the following day. Hence the recoinage was both decided and inaugurated well before Newton's appointment as warden of the Mint. Nor did his opinion on the recoinage determine the policy, which differed in a minor way from what he recommended. In no sense did Newton bear responsibility for the recoinage. He did accept responsibility for carrying it through to completion.

Montague dated his letter offering the appointment 19 March. Newton went to London immediately. The Exit and Redit Book shows that he left on 23 March, and we know he was in London on 25 March. He did not spend long agonizing over his decision; the warrant for his appointment was drawn up that same day. He had packed his belongings by 20 April, when he left Trinity for good. It took him nearly as long to settle in in London as he had taken to leave Cambridge, but on 2 May he was ready to swear the special oath demanded of all personnel at the Mint. Four days later, in conjunction with Thomas Neale and Thomas Hall, the master and his assistant in charge of the recoinage, he signed his first official communication to the Treasury.

Newton's decision to engage himself closely in the recoinage was his own free choice of a duty not necessarily incumbent on the warden. Under the new constitution of the Mint, which had been instituted in 1666, the master and worker had become the true authority. Newton's predecessors in the office of warden had treated it as a sinecure.

Years later, the Earl of Halifax (as Montague became) used to remark that he could not have carried on the recoinage without Newton. At the time, however, Montague offered the position as the sinecure it had become. It was worth, he wrote, five to six hundred pounds (a deliberate exaggeration) "and has not too much bus'nesse to require more attendance then you may spare." We can only speculate on Montague's motives. He had become Newton's friend in Trinity College, where his cousin was then master, in the 1680s, and he may have been merely following the desire to satisfy a friend. The politics of revolutionary England did not ordinarily operate on that friendly basis, however. Patronage was the very marrow of power. Montague had only recently arrived at a position of power, and it is unlikely that he would have expended such a ripe plum as the wardenship of the Mint wantonly. The Whig Junto was known for quite the opposite. What advantage could Montague have expected to extract from the appointment of Newton? Most wardens sat in Parliament and with other holders of similar places supported the government. In Newton's case at least, we need not pitch the expectations on a servile plane. He had made his opinions known. He had also stood for Parliament successfully on one occasion. The Steward's Book at Trinity showed that he spent half a week there in 1698, and his ballot in that year's election survives. We may speculate that he canvassed the possibility of standing himself in future elections. At least he did so in the next election, in 1701, when he was returned. He considered it strongly in 1702 and stood once more in 1705, when his defeat wrote finis to his Parliamentary career.

When Montague offered a position that would not absorb much of his time, he did not take Newton's need to escape from intellectual activity into account, or his inability to do anything halfway. From the first, Newton flung himself fully into the recoinage. Thomas Neale, the master, a political adventurer who dabbled in any enterprise likely to return profit, from the postal service in the American colonies to the lotteries designed to make war taxes more palatable, was too distracted to give the recoinage the attention it required, and when Newton arrived the recoinage was floundering, compounding the crises of 1696 as the strain of the war stretched the revolutionary government almost to the breaking point. The Mint needed all the energetic and intelligent leadership it could find. Newton gave it his all. It is instructive that Halley, for whom Newton obtained a post in the Chester mint, complained that the work was drudgery at best. Not only did Newton never make a similar complaint, but he contrived to make his position, which for others was temporary, into a permanent one.

In the memoir that he wrote on the recoinage, Hopton Haynes, a clerk at the Mint whose patron Newton became, remarked that Newton's skill in numbers enabled him to comprehend the Mint's accounting system immediately. Undoubtedly Haynes was correct, but the gifts that Newton brought to the position were not confined to understanding accounts. He possessed an innate tendency to order and to categorize. His first step in every new intellectual undertaking was the composition of some sort of index to help him organize his knowledge. The same tendency served him well at the Mint. In the perspective of history, the recoinage appears a mean thing in comparison to the *Principia*. Be that as it may, Newton had made his choice. He was a born administrator, and the Mint felt the benefit of his presence.

One aspect of the recoinage that had not progressed well was the erection of five temporary country mints to speed the diffusion of the new coin throughout the realm. When he arrived, the country mints were badly behind schedule, and the Treasury was pressing hard for them. They were one of the tasks to which he turned his hand. The extent of Newton's responsibility for their successful operation cannot be demonstrated precisely; evidence that he played a role consists primarily of chronological coincidence. In fact they did begin to function less than three months after he arrived, although he was not the only person at the Mint involved.

As a result of his energy, he found himself in a position finally to repay some of his debt to Halley. Each of the country mints needed its full delegation of officers, who were appointed as deputies of the officers of the Tower Mint. Newton arranged for Halley's appointment as the deputy comptroller in the Chester mint, at a salary of £90 per annum. As it turned out for Halley, headaches worth twice that sum accompanied the post.

Within the Mint at the Tower, all was frantic activity. Lord Lucas, the governor of the Tower, thought that five o'clock in the morning was early enough to open the gates; the Treasury ordered him to open them at four o'clock. Work continued until midnight. According to Hopton Haynes, who wrote a historical memoir of the recoinage, nearly three hundred workmen crowded into the narrow confines of the Mint, and fifty horses turned the ten mills that operated. Nine great presses worked, each striking according to Newton's calculation between fifty and fifty-five times a minute with what must have been an incredible din. By heroic efforts, the Mint managed to push its production up to £100,000 per week during the summer of 1696, and by the end of the year it had coined £2,500,000. By that time, as the monetary shortage began to ease, the worst of the crisis had passed. Haynes attributed considerable credit for the performance to

Newton. Because Newton appears to have commissioned Haynes's *Brief Memoires*, we should treat his testimony with caution. Nevertheless, the warden did fling himself into the task with immense vigor.

It did not take Newton long to size up realities in the Mint. He had accepted the position under the impression that the warden held the highest authority. By June he understood otherwise and petitioned the Treasury for an increase in salary. The warden's salary, he complained, "is so small in respect of the Salaries & Perquisites of the other Officers of the Mint as suffices not to support the authority of his Office." He sized up the master, Neale, as well − "a Gentleman who was in debt & of a prodigal temper & by irregular practices insinuated himself into ye Office [of master] . . . ," as he remarked later. Tension between them had developed by the time of a parliamentary investigation in 1697. In the papers that he composed for the committee of the Commons, Newton tried again to reassert the authority of the warden: "The Warden is . . . by his Office a Magistrate & the only Magistrate set over the Mints to do Justice amongst the members thereof in all things . . . . The Workers (one of wch is Master of ye rest) are they who melt refine allay & run the standared gold & silver into Ingots to be coyned." Originally, no workers were standing officers, but the reorganization of 1666 gave the master a salary higher than the warden's and appointed him to receive and distribute the income of the Mint from the coinage duty. The power of the purse brought the Mint under the master's control. In his report Newton urged the restoration of the earlier state of affairs.

The attempt to remodel the constitution of the Mint never got into motion. Adopting an alternative strategy, Newton set about making himself master in fact if not in name. By the methods he had applied in enterprises of a wholly different nature, he undertook a systematic study both of the history of the Mint and of its present operation so that he might stand on a footing of incontestable knowledge. He collected copies of proclamations and warrants relevant to the Mint stretching back to the reign of Edward IV in the fifteenth century. He took care to inform himself of Neale's affairs, especially his indebtedness to James Hoare, a former comptroller of the Mint. He pored over the old accounts to become familiar with the level of payments for various services. He studied each of the operations of the Mint in detail, recording the various expenses it involved, such as the cost of a melting pot and the number of times it could be used. "By experim^t I found that a pound Troy of ½ crown blancks lost 3½ gr. [in blanching]," he noted.

One fascinating aspect of Newton's habits emerges from these papers. It remained characteristic of all his papers at the Mint and helps to illuminate his other papers, among which multiple drafts are very common. Newton was an obsessive copier. The Newtonian scholars A. R. and Marie Boas Hall have suggested that he could not read attentively without a pen in his hand. By confirming this habit with material that required no creative thought, the Mint papers serve to underscore it. With a stable of amanuenses at his command, Newton copied a report of 1675 on the state of the coinage, and then copied it a second time. He copied the record of the amount coined, both by weight and by tale, both in gold and in silver, year by year from 1659 to 1691 – and again copied it all a second time. In part, the copying stemmed from the conviction that he could rely with confidence on himself alone. As he advised the officers of the country mints in the instructions for their accounts, "trust not the computation of a single Clerk nor any other eyes then your own." The matter went beyond trust, however. Even a minor letter could extract two drafts and two fair copies from him.

By the completion of the recoinage in the summer of 1698, Newton had mastered the operation of the Mint to the extent that he had virtually assumed the title of master. The lord commissioners asked him to take charge of drawing up the final accounts of the country mints, which Neale would ordinarily have done. He also performed Neale's task of composing the final report of the recoinage. Indeed, Neale's records stood in such impossible shape that he did not succeed in clearing his accounts before his death at the end of 1699, and they remained an additional task for Newton – a symbol of the true situation within the Mint well before his death.

According to Newton's records, the Mint (including the country mints) succeeded in recoining £6.8 million from the beginning of 1696 to the summer of 1698 – nearly twice the total coinage, measured by the number of coins, of the previous thirty years. Through no fault of his, it all came to naught. Despite the government's willingness to impose a sudden deflation on an economy strained to the breaking point by the war and to risk social upheaval by its inequitable provisions, it did nothing to correct the basic undervaluation of silver. Almost as fast as it issued from the Mint, the new coinage went into the melting pots of the goldsmiths. During the rest of Newton's life, the Mint coined silver only at those times when special acts of the government brought plate and bullion into the Mint, and even then it coined very little. Within two decades, the Mint was experimenting with quarter-guineas, that is, gold coins, to relieve the shortage in small denominations. John Conduitt, the husband of Newton's niece and his successor at the Mint, noted in his "Observations" on the coin in 1730 that very little silver remained in circulation.

෧ঌ

The office of warden involved one dimension for which Newton may not have bargained initially. The warden was charged with the apprehension and prosecution of counterfeiters. Newton's first reaction was to shrink from the job. Apparently in the summer of 1696, he wrote to the Treasury requesting that a duty "so vexatious & dangerous" not be required of him. The job belonged more properly to the Solicitor General. The Treasury decided to proceed in the opposite direction by authorizing additional funds for the employment of another clerk.

Unable to slough the obligation off, Newton plunged into it with customary thoroughness. Professor Manuel has argued that Newton's prosecution of coiners gave expression to his suppressed aggressions and that in coiners he found socially acceptable objects on which vicariously to wreak vengeance on his stepfather. In the context both of the times and of Newton's career at the Mint, his prosecution of coiners looks a great deal less peculiar. He served as warden at the time of maximum concern about the debasement of the currency. New legislation of the 1690s and immense attention to the problem before his appointment make it clear that Newton did not invent the pursuit of coiners by a quirk of his own tortured psyche. Moreover, the prosecution of coiners continued under other wardens after Newton's elevation to the mastership though less and less connected with the warden himself because it was increasingly professionalized in the hands of his special assistants.

It is certainly true, as Professor Manuel contends, that Newton invested great energy in the task. The conviction of a coiner required witnesses. Newton was industry itself in finding them. The surviving records contain as many as fifty-eight depositions before him in the space of two months, depositions taken in taverns and in the less salubrious atmosphere of Newgate and other prisons in London. Among his correspondence, some of the endless letters required to transfer prisoners from jail to jail in order to have them present in court survive, only an indication of the much larger number he must have written. He had himself commissioned as a justice of the peace in all of the home counties. His accounts showed the operation of agents in eleven counties. He bought special clothes for Humphrey Hall "to qualify him for conversing with a Gang of coyners of Note in order to discover them." The accounts named twenty-eight coiners whom he prosecuted successfully. In addition they included several payments for the discovery of coiners (in the plural) plus one payment to a man who prosecuted twenty-six persons in addition to seventeen whom he convicted. With the additional names that appear in the Treasury Books, Treasury Papers, and State Papers, there is evidence for the pursuit of far more than a hundred coiners, not all by Newton personally, of course.

Of all the coiners, none was more colorful or more resourceful than William Chaloner, one of those whom the lord justices sent to Newton for questioning in the summer of 1696. Before he had done with Chaloner, Newton had collected most of his life history, which he set out in a memorandum for Parliament. "A japanner in clothes threadbare, ragged, and daubed with colours," Chaloner "turned coiner and in a short time put on the habit of a gentleman." He took up his new trade about 1690, working at first mostly in foreign coins, which passed continually in the daily life of London. An artist among counterfeiters, he was the author of a new method of coining which Newton found the most dangerous he had encountered. He also had flair; and, alone among the ominous and distasteful denizens of the coining world, Chaloner saw the possibilities in playing both sides of the street. He started, not with coinage, but with Jacobite propaganda. In the early 1690s, he cajoled a couple of printers into producing some declarations in favor of James and then turned them in for a reward of £1,000. In his own private terminology, which even his associates had to interpret, he "funned" (that is, deceived) the king of £1,000. Impressed by the ease of his gain, he discovered a plot to cheat the Bank and funned it of £200. Unfortunately for Chaloner, he lacked the sense to realize that he could not play the same act to the same audience forever. In February 1696, just before Newton arrived, he tried it a third time by submitting two papers to the Privy Council on abuses in the Mint and methods of preventing counterfeiting. Shortly thereafter, he made the acquaintance of Newton by suggesting one of his associates, Thomas Holloway, as a man suitable to be his special clerk to pursue coiners. All was not merry deception of the establishment, however. The unfortunate printers went to the gallows, and Chaloner contrived to have two coining associates "hanged out of the way" when they informed on him under duress.

Newton's serious concern with Chaloner began in 1697 when he decided to fun the Parliament. Chaloner testified to the committee investigating abuses in the Mint that he could improve the coinage at no extra cost in a way that would prevent counterfeiting. His imagination now running riot, he proposed that he be installed as supervisor of the Mint to oversee his improvements. He also thought he would fun the government once more with the Jacobite gambit. It was once too often. Skeptical lord justices heard his charges in June; they sought further information. In August, they heard Newton's testimony of Chaloner's ongoing enterprise in coining; and early in September, they charged Newton not to let Chaloner, then in prison, be bailed. Had Chaloner known that the highest council of the government was hearing his case on a regular basis during the summer of 1697 and expressing its desire to execute him for treason if it could collect enough evidence to convict him, he might have chosen to lie low. Not knowing, he rushed on to his fate. Ever audacious, he petitioned Parliament, charging that the Mint was trying to destroy him in revenge for this testimony against them the previous session. The committee appointed to investigate the petition included Secretary Vernon, Montague, and Lowndes, who were all aware of Chaloner's activities. Chaloner was not aware of theirs.

He had one last act to play, for he succeeded in gaining his release from prison sometime early in 1698. As Newton later learned by diligent enquiry, he bought off the chief witness against him, Thomas Holloway, and got him to flee to Scotland. One Henry Saunders told Newton that when he visited Chaloner in Newgate to tell him Holloway had gone, he "seemed to be very joyfull and said a fart for ye world." As soon as he was released, he organized or joined a flatulent enterprise to forge malt tickets, one of the new devices of paper currency, issued in connection with a duty on malt enacted the previous year. Already in May, Montague and

Secretary Vernon began to receive evidence about the plan. It was not Newton but Vernon, who had become convinced the previous autumn that Chaloner was too dangerous to be at large, who engineered his doom and issued the final warrant for his arrest. It was Newton, however, who sealed his fate by weaving a net of evidence about him from which he could not escape. When Holloway returned to London from Scotland, Vernon sent news of him to Newton, who took him into custody. Already in 1698, in one draft of his memorandum to Parliament on Chaloner's petition, Newton cited fourteen witnesses against him. In late 1698 and early 1699, he took more than thirty further depositions as he pieced together the story both of the malt-ticket caper and of earlier activities. When Chaloner returned to Newgate, Newton constructed a circle of spies to inform him of Chaloner's every stratagem.

Despite a bevy of lawyers in daily attendence and a pretense of madness, Chaloner was convicted of high treason on 3 March 1699. He had cut a sufficient figure and could afford lawyers good enough that the king himself heard his petition for pardon on 17 March. As he faced the terrible punishment that bloody imaginations had devised for high treason, he finally collapsed.

Most mercifull S[r]

I am going to be murther[d] allthough perhaps you may think not but tis true I shall be murdered the worst of all murders that is in the face of Justice unless I am rescued by yo[r] mercifull hands.

S[r]

pray considr my unprecedented Tryall 1 That no person swore they ever saw me actually coyn y[t] I should own it . . . 6[ly] It was hard for me to be taken out of my 5 weeks Sick Bedd the last 3 weeks light headed So y[t] I was not provided for a tryall nor in my Senses w[n] tryd 7[ly] W[t] Mrs Carter swore ag[t] may appear direct mallice I have 3 yeares before Convicted her husband of Forgery and discovered where he and she were coyning for w[ch] he is now in Newgate But I desire God Allmighty may Damne my Soul to eternity if every word was not false that Mrs Carter and her Maid swore ag[t] me abo[t] coyning and Mault Tickets for I never had any thing to do with her in coyning nor ever intended to be concerned in Mault Tick[ts] nor ever spoke to her abo[t] any such things M[rs] Holloway swore false ag[t] me or I desire never to see the Great God and I desire the same if Abbot did not swear false ag[t] me so y[t] I am murder[d] O God Allmighty knows I am murder[d] Therefore I humbly begg yo[r] Wor[p] will consid[r] these Reasons and y[t] I am convicted without Preced[ts] and be pleased to speak to M[r] Chancell[r] to save me from being murtherd O Dear

S^r do this mercifull deed O my offending you has brought this upon me O for Gods sake if not mine Keep me from being murder^d O dear S^r no body can save me but you O God my God I shall be murder^d unless you save me O I hope God will move yor heart with mercy and pitty to do this thing for me I am

yor near murder^d humble Servt
W. Chaloner

It wasn't possible. The government that existed only for Chaloner to fun wasn't really going to execute its unthinkable sentence on him. In fact it was, and Newton could not have stayed the remorseless machine of justice had he wished.

Thursday, 23 March [Narcissus Luttrell recorded in his *Historical Relation*]. Yesterday 7 of the criminals, condemned last sessions at the Old Baily, were executed at Tyburn; Challoner, for coining, drawn on a sledge; Mr. John Arthur, for robbing the mail, was carried in a coach; and 5 other men for robbery and burglary.

Further entertainment awaited Chaloner at Tyburn before death put an end to his suffering and to his career in coining.

❦

Most of what we know about Newton's daily life in London comes from a later time – a few miscellaneous bills, the inventory of his goods and chattels after his death, the bills paid by his estate, and the comments of John Conduitt, who lived with him for a number of years. Although we should exercise some caution in using them, there is no reason to think that Newton later changed his habits in any marked degree. Conduitt's summary fits the evidence admirably. "He always lived in a very handsome generous manner tho^u without ostentation or vanity, always hospitable & upon proper occasions gave splendid entertainments." That is, without attempting magnificence, he lived in a style consistent with his new dignity. In summarizing the inventory of his goods, Richard de Villamil, who published Newton's will, contrived to paint a picture of spartan utilitarianism in his household furnishings. The inventory showed a well-furnished house, however, and I do not see how one can judge the quality and artistry of furniture, as Villamil did, from bare descriptions as tables and chairs and the like. One surviving bill recorded the purchase of four landscapes to decorate the walls and twelve Delft plates. The inventory, with three dishes, three salvers, a coffeepot, and two candlesticks (all of silver), forty plates, a full set of silver flatware, about ten dozen glasses and six-and-a-half-dozen napkins, demonstrated that he had the equipment for the occasional splendid entertainments that Conduitt mentioned. For other needs, he possessed no fewer than two silver chamberpots, which no one would characterize as spartan utilitarianism. He owned clothes valued at only £8 3s 0d; but by the time of the inventory, Newton had for five years been a semi-invalid afflicted with incontinence of urine, which would take its toll of any wardrobe. As Villamil remarked, he had a penchant for crimson – crimson draperies, a crimson mohair bed with crimson curtains, crimson hangings, a crimson settee. Crimson was the only color mentioned in the inventory, and Villamil suggested with justice that he lived in an "atmosphere of crimson." Before his late illness, he apparently kept a coach; and he maintained a stable of servants, six at the time of his death. Years before, Newton had resented his servile status as a sizar. There is every reason to believe that he now seized the opportunity to adopt the style of the better circles of London society and that he relished doing so.

As for his table, Conduitt reports that he was always very temperate in his diet. One note spoke of his living on vegetables, though another denied that he abstained from meat. Perhaps Conduitt's information was consistent with the judgment of the Abbé Alari, the instructor of Louis XV, who dined with him in 1725 and found the repast detestable. He complained that Newton was stingy and served poor wines which had been given to him as presents. Because French visitors invariably commented on English cuisine in a similar vein, we cannot conclude too much from Alari's dyspepsia. A bill showing the delivery of one goose, two turkeys, two rabbits, and one chicken to the household in the space of a single week reminds us that Conduitt was applying eighteenth-century standards when he described Newton's diet as temperate. After his death, his estate settled a debt of £10 16s 4d with a butcher and two others, which totaled £2 8s 9d, with a poulterer and a fishmonger. In contrast, he owed the "fruiter" only 19s and the grocer £2 8s 5d. A bill of £7 10s 0d for fifteen barrels of beer again suggests less than heroic temperance.

The move to London did not alter his habits, especially his penchant for constant study. His former studies in turn refused to leave him alone. On 29 January 1697, he received two challenge problems issued by Johann Bernoulli. Bernoulli had published one of them originally in the *Acta eruditorum* the previous June – to find the path by which a heavy body will descend most quickly from one point to another that is not directly beneath it – and he had set a limit of six months to the challenge. When December came, he had not yet received a satisfactory answer, though he had received a letter from Leibniz with both the assertion that he had solved the problem and the request that the time be extended to Easter and the problem republished throughout Europe. In accepting Leibniz's request, Bernoulli added a second problem. He had copies of the problems sent to the *Philosophical Transactions* and the *Journal des sçavans*. He also sent copies to Wallis and Newton. Recall that earlier in 1696 Bernoulli had expressed the opinion that Newton had filched from Leibniz's papers the method that he first published in Wallis's *Opera*. Manifestly, both Bernoulli and Leibniz interpreted the silence from June to December to mean that the problem had baffled Newton. They intended now to demonstrate their superiority publicly. In case the direct mailing to him were not pointed enough, Bernoulli inserted a scarcely veiled reference in the announcement itself. He and Leibniz would publish their solutions at Easter, he stated.

If geometers carefully examine these solutions, drawn as they are from what may be called a deeper well, we are in no doubt but that they will recognize the narrow limits of the common geometry, and will value our discoveries so much the more as there are fewer who are likely to solve our excellent problems, aye, fewer even among the very mathematicians who boast that by the remarkable methods they so greatly commend, they have not only penetrated deeply the secret places of esoteric geometry but have also wonderfully extended its bounds by means of the golden theorems which (they thought) were known to no one, but which in fact had long previously been published by others.

Whatever Bernoulli – and Leibniz – had in mind, Newton saw the problems as a challenge issued personally to him. He accepted the challenge by recording on the paper the time at which it arrived. "I received the sheet from France, Jan. 29. 1696/7." He dated a letter to Charles Montague, president of the Royal Society, which set down the answers to both problems, 30 January. His sense of triumph was great enough that the story made its way, via his niece Catherine, into Conduitt's collection of anecdotes. "When the problem in 1697 was sent by Bernoulli – Sr I. N. was in the midst of the hurry of the great recoinage did not come home till four from the Tower very much tired, but did not sleep till he had solved it wch was by 4 in the morning." In addition to Leibniz's solution,

Bernoulli received two others, one from the Marquis de l'Hôpital in France and an anonymous one from England. Disabused on Newton's skill in mathematics, Bernoulli recognized the author in the authority the paper displayed – "as the lion is recognized from his paw," in his classic phrase.

Even without the stimulus of a challenge problem, Newton also allowed the moon to occupy him somewhat. Already by September 1697, a letter from Flamsteed indicated that the two had discussed the moon in London. In December of the following year, Newton visited Greenwich, and Flamsteed began again to supply him with data. Their relations came to a head once more at the end of 1698. Under constant pressure to justify himself by publishing, Flamsteed yielded that year to Wallis's entreaty that he print an account of his supposed observation of stellar parallax in the final volume of Wallis's *Opera*, which was due to appear in 1699. As he put the matter later to Newton, he used his account "to silence some busy people y^t are allwayes askeing, *why I did not print?*" Hence he recited his accomplishments as Astronomer Royal, a new catalogue of the fixed stars, for example, and rectified solar tables. Assailed as he was by repeated rumors, which he could not reasonably doubt, of Newton's complaints in regard to lunar observations – including one story that Newton claimed to have rectified the lunar theory with Halley's observations – Flamsteed also added a paragraph on what he had done in that respect.

I had also become closely associated with the very learned Newton (at that time the very learned Professor of Mathematics in the University of Cambridge) to whom I had given 150 places of the Moon, deduced from my earlier observations and her places at the times of the observations as computed from my tables, and I had promised him similar ones for the future as I obtained them, together with my calculations, for the purpose of improving the Horroxian theory of the Moon, in which matter I hope he will have success comparable to his expectations.

Because Wallis's colleague David Gregory was in London, Flamsteed sent the paper to Wallis via him. On 31 December he received a letter from Wallis. Wallis had heard from an unnamed correspondent in London, a friend both of Newton and of Flamsteed, who asked Wallis without stating his reasons not to print the paragraph about Newton. Flamsteed recognized Gregory's work and noted on the letter that he was clearly no friend of his. He wrote to Newton at once.

S^r My observations lie ye King & Nation in at least 5000 *lib*, I have spent above 1000 *lib*. out of my own pocket in building Instrum^{ts} & hireing a servant to assist me now neare 24 yeares; tis time for me (& I am now ready for it) to let the World see I have done something that may answer this expence. & therefore I hope you will not envy me the honor of haveing said I have been usefull to you in your attempts to restore the Theory of the Moon I might have added the Observations of the Comets places given you formerly of the superior planets & refractions at ye same time wth y^e ☽s [moons]. but this I thought would look like boasting & therefore forbore it.

He received no response. He wrote a second time and again heard nothing. On 7 January, he wrote to Wallis that Newton apparently was unconcerned and that the paragraph might stand.

He wrote too soon. Newton's letter, dated 6 January, arrived as soon as he mailed his own to Wallis. In it, Newton cast aside the mask of friendship and assaulted Flamsteed with all the brutality of which he was capable.

S^r

Upon hearing occasionally that you had sent a letter to Dr Wallis about ye Parallax of ye fixt starrs to be printed & that you had mentioned me therein with respect to ye Theory of ye Moon I was concerned to be publickly brought upon ye stage about what perhaps will never be fitted for ye publick & thereby the world put into an expectation of what perhaps they are never like to have. I do not love to be printed upon every occasion much less to be dunned & teezed by forreigners about Mathematical things or to be thought by our own people to be trifling away my time about them when I should be about ye Kings business. And therefore I desired Dr Gregory to write to Dr Wallis against printing that clause wch related to that Theory & mentioned me about it. You may let the world know if you please how well you are stored wth observations of all sorts & what calculations you have made towards rectifying the Theories of ye heavenly motions: But there may be cases wherein your friends should not be published without their leave. And therefore I hope you will so order the matter that I may not on this occasion be brought upon the stage. I am

Your humble servant
Is. Newton

In his reply, Flamsteed pointedly mentioned Newton's own willingness to advertise his lunar theory orally and took exception to the implication that his own work, and Newton's as well, was trifling. But he bowed to the inevitable and told Wallis to remove "ye Offensive Innocent Paragraph. . . ."

§

By this time, Newton was a famous man, and visitors to London from abroad who were informed about natural philosophy made it their business to meet him. No visit held more significance than that of Jacques Cassini in the spring of 1698. According to Conduitt, who must have repeated what he heard from Newton much later, Cassini offered Newton a large pension from Louis XIV. This could have referred only to an appointment in the Academy of Science, which was then being reorganized. Newton declined. The reorganization also established eight foreign associates, of which the King appointed three and the academy elected five. If Conduitt's story was correct, we can understand why Louis chose not to include Newton among his nominees of foreign associates (Leibniz, Ehrenfried von Tschirnhaus, and the relatively obscure Italian physicist, Domenico Guglielmini). Newton did accept election by the academy, however, along with Nicolas Hartsoeker, Ole Roemer, and the two Bernoullis.

Although after 1693 Newton never resumed a correspondence with Locke as intimate as their earlier one, the two did remain in contact. In the autumn of 1702, Newton visited Oates and saw there Locke's newly completed commentary on 1 and 2 Corinthians. Since he did not have time to study it carefully, he asked Locke to send him a copy. When Locke heard nothing in reply, he wrote in March. Still awaiting a reply, he wrote a second time late in April and sent the letter to his cousin Peter King with the request that he deliver the note in person.

The reason why I desire you to deliver it to him yourself is, that I would fain discover the reason of his so long silence. I have several reasons to think him truly my friend, but he is a nice man to deal with, and a little too apt to raise in himself suspicions where there is no ground; therefore, when you talk to him of my papers, and of his opinion of them, pray do it with all the tenderness in the world, and discover, if you can, why he kept them so long, and was so silent. But this you must do without asking him why he did so, or discovering in the least that you are desirous to know. . . . Mr. Newton is really a very valuable man, not only for his wonderful skill in mathematics, but in divinity too, and his great knowledge in the Scriptures, wherein I know few his equals. And therefore pray manage the whole matter so as not only to preserve me in his good opinion, but to increase me in it; and be sure to press him to nothing, but what he is forward in himself to do.

Stimulated by King's visit, Newton finally wrote on 15 May, apologizing for his long silence and commenting at length on 1 Corinthians 7:14, on the meaning of which he differed from Locke. Apparently the two never met again.

Somewhat unexpectedly, Newton also found theological companionship at the Mint in the person of Hopton Haynes. He set Haynes to work composing a history of the recoinage. The resultant *Brief Memoires of the Recoinage,* a panegyric to Newton, drew heavily on material found among Newton's papers, which he must have furnished. Newton did not hesitate to ask for favors in return. In 1701, he was working on a report on foreign coins and needed a copy of an earlier one prepared in 1692, which was available in the Excise office. In the accepted practice of the day, Haynes held a clerkship there as well, and Newton got him to copy it. To the copy, Haynes appended a personal note which helps to clarify their relation.

I am sorry I had not ye good fortune to see You ysday at the Excise, tho' I hope You are so kind, as stil to continue y^r good inclination to favour my pretension, if occasion be.

But I have rec^d such demonstration of y^r friendship, allready, for which I never pretend to return, & You, I dare say, never expect any other requital than my gratitude that I cannot but assure myself of Y^r good Offices when a fair occasion presents, by which You'll extremely add to the many obligations You have already layd upon

S^r Y^r most Obed^{nt}
& most H. Servt
H. Haynes

A fair occasion did in fact present itself the following year when the post of weigher and teller fell empty. Newton favored Haynes's pretensions to the extent of composing six successive drafts of his recommendations. Needless to say, Haynes became the weigher and teller at the Mint; he held the position until 1723, when Newton secured his appointment as assaymaster. In 1714, Newton consulted Haynes about the design of the coronation medal for George I. Haynes married about 1698, and like Humphrey Newton, he named his fourth son, born sometime near 1705, after Newton.

What we know of their theological relationship is confined to Haynes's assertions and to reports of Haynes's assertions. Richard Baron, an abrasive unitarian who described Haynes years later as "the most zealous unitarian" he had ever known, reported that Haynes had told him Newton held the same views. From the time of the Convention Parliament, Newton had discovered the possibility of discreet discourse in London on matters left untouched in Cambridge. With Locke, with Fatio, with Halley, and with Bentley, at various times and in various ways, he exchanged honest theological opinions. Near the time of his move, he apparently did so as well with a young man in Cambridge, William Whiston. Without venturing to imagine the circumstances, we must assume that Newton early recognized in Haynes one who held similar views, or was capable of holding them. It seems more than mere speculation that Newton's patronage of Haynes, like his patronage of Whiston, rested in good part on theological agreement. It also seems more than mere speculation that both young men learned the greater portion of their heresy from Newton.

In public Newton chose to disguise his heterodoxy. He allowed himself to be appointed as a trustee of the Golden Square Tabernacle, a chapel endowed by Archbishop Tenison to relieve the crowding at St. James's, in the parish of which Newton's house on Jermyn Street stood. He also became a member of the commission to complete St. Paul's until he disputed one day with Archbishop Wake about hanging pictures in the cathedral. Newton related that Wake "told a story of a Bishop who said on that subject that when this snow (pointing to his grey hairs) falls, there will be a great deal of dirt in churches . . . ." According to Catherine Conduitt, he never attended another meeting of the commission.

è&

Catherine Conduitt, who supplied the story of the bishop and much else to her husband, John Conduitt, was Newton's niece, née Catherine Barton. According to Conduitt, she lived with Newton twenty years before and after her marriage. The daughter of Newton's half-sister Hannah Smith, who married Robert Barton, a cleric in Northamptonshire, Catherine was born in 1679. Her father died in 1693, leaving her mother nearly destitute, if we can judge by Hannah Barton's letter to Newton at the time; and Newton purchased an annuity for her three children around 1695. Once he settled into the house on Jermyn Street, he made arrangements for Catherine to live with him. No evidence establishes the time at which she joined Newton, though a letter that Newton addressed to her in

August 1700 seems to suggest by its tone that she had lived with him already some time. By every account, Catherine Barton possessed unlimited charm and was a woman of beauty and wit. She was the one member of Newton's family who seemed to share his talents, though being a woman she had to exercise hers in rather different channels. In eighteenth-century terminology, Newton's niece was the famous witty Mrs. Barton.

When Voltaire visited England in the 1720s, he too heard of Catherine Barton, and what Voltaire heard, all of Europe heard.

I thought in my youth that Newton made his fortune by his merit. I supposed that the Court and the city of London named him Master of the Mint by acclamation. No such thing. Isaac Newton had a very charming niece, Madame Conduitt, who made a conquest of the minister Halifax. Fluxions and gravitation would have been of no use without a pretty niece.

The story was not invented by Voltaire, and it was not fabricated wholly from someone's imagination. By 1703, if not before, Halifax (as Montague now styled himself) had made Catherine Barton's acquaintance, and in 1706 he drew up his will. Two days after doing so, he added a codicil which bequeathed £3,000 and all his jewels to Catherine Barton "as a small Token of the great Love and Affection I have long had for her." In October 1706, he added to the bequest by purchasing for her, in the name of Isaac Newton, an annuity of £200 per annum for the remainder of her life.

The official life of Halifax commissioned by his heir and published soon after his death felt obliged to mention their relation.

I am likewise to account for another Omission in the Course of this History, which is that of the Death of the Lord Halifax's Lady; upon whose Decease, his Lordship took a Resolution of living single thence forward, and cast his Eye upon the Widow [rather, sister] of one Colonel Barton, and Neice of the famous Sir Isaac Newton, to be Super-intendant of his domestick Affairs. But as this Lady was young, beautiful, and gay, so those that were given to censure, pass'd a Judgment upon her which she no Ways merited, since she was a Woman of strict Honour and Virtue; and tho' she might be agreeable to his Lordship in every Particular, that noble Peer's Complaisance to her, proceeded wholly from the great Esteem he had for her Wit and most exquisite Understanding . . . .

Three thousand pounds and all his jewels plus the annuity seems to most a stiffer price than wit and understanding normally command.

On 1 February 1713, Halifax drew up a second codicil to his will which revoked the first and replaced it with one nothing short of magnificent. To Isaac Newton he left £100 "as a Mark of the great Honour and Esteem I have for so Great a Man." To Newton's niece, called here Mrs. Catherine Barton, he bequeathed £5,000 with the grant during her life of the rangership and lodge of Bushey Park (a royal park immediately north of Hampton Court) and all its furnishings and, to enable her to maintain the house and garden, the manor of Apscourt in Surrey. "These Gifts and Legacies, I leave to her as a Token of the sincere Love, Affection, and Esteem I have long had for her Person, and as a small Recompence for the Pleasure and Happiness I have had in her Conversation." When Flamsteed heard about the bequest after Halifax's death, he spitefully wrote to Abraham Sharp that it was given Mrs. Barton "for her *excellent conversation*." Flamsteed placed a value of £20,000 or more on the house and lands – that is, he valued the whole bequest at £25,000 or more plus the annuity, a fortune by the standards of the early eighteenth century. Flamsteed also remarked that Halifax was reported to have left a total estate of £150,000, handsome testimony to what an enterprising lad could accomplish with a mere five years in office.

The problem with the codicil of 1713 is sheer size. If the bequest of 1706 makes it impossible to believe in a platonic relation between Halifax and Mrs. Barton, the later one makes it difficult to believe she was merely a mistress. If not a mistress, what? If there had been a secret marriage, as some have argued, why was it kept secret? The motive proposed – Halifax's fear of ridicule for marrying beneath his class – carries no conviction. Even if we allow it, why conceal the marriage after his death, when the bequest left his supposed widow exposed to slander? When Catherine Barton married John Conduitt in 1717, she recorded herself in the presence of her groom, who could not have been ignorant of her liaison with Halifax or of its material legacy, as a spinster. Because there are reasons to doubt that she was either wife or mistress, perhaps she occupied some intermediate status. If such exists, I have not heard of it.

The issue here is Newton's role in and attitude toward the affair, which began in the early years of his residence in London. It has been felt that his acquiescence in his niece's relation to Halifax, which was clearly not a legal marriage, must somehow diminish his stature. Merely to pose the issue appears to me to assume that Newton stood on a moral plane separate from and superior to the society in which he lived. And yet, for all his genius, he was a human being like all of us, facing similar moral choices in terms not altered by his intellectual achievement. His success at the Mint – in administering the recoinage, in dealing with the gold-smiths, above all in maneuvering himself into position to become master when he realized the wardenship was a sham – does not suggest an otherworldly saintliness out of touch with hard reality. He knew what compromise was. His pretense of religious conformity for social accep-tance and material benefit was not utterly incommensurable with acquies-cence in a most advantageous liaison. For that matter, he knew what sexual attraction was – and from every indication, its gratification, of necessity outside the bonds of holy wedlock. Newton's role in history was intellectual, not moral, leadership. From the point of view of the late twentieth century, after the barbarities we have witnessed, the charge against him does not seem oppressively heavy, but even if it could be proved beyond doubt that Newton was the leading whoremonger of Lon-don, the immensity of his impact on the modern intellect would remain unaltered. For me at least, the recognition of his complexity as a man helps in understanding the price his genius exacted. I find it hard to reconcile the *Principia* with a plaster saint.

On 23 December 1699, Thomas Neale, the master of the Mint, died. It had not taken Newton long to understand realities at the Mint, and it had not taken him much longer to realize that his effort to reverse the shift of true authority from the warden to the master was not going to float. As he observed the recoinage, the disparity between the seats of formal and real authority, not to mention their disparity in remuneration, must have struck him more sharply. Although he shouldered great burdens in the recoinage, he received the same salary, £400 per annum, he would have received had he acted like former wardens and done nothing. Neale did very little, leaving all to his assistant, Thomas Hall, and to Newton. Not only did Neale receive his salary of £500 per annum, but he got, accord-ing to the terms of his Indenture, a set profit on every pound weight troy that was coined. In addition to his salary, Neale earned more than £22,000 during the recoinage. Newton perceived and digested this fact. He schooled himself in the operations of the Mint that a master should know. And he waited, for Neale was a sick and failing man.

At first glance, it appears that Neale lasted too long, for Montague had fallen from power by the time he died. We can only speculate on why it did not matter. As Neale's profits on the recoinage indicated, the master-ship could be a considerable item of patronage. Nevertheless, even with Montague out of office, Newton was allowed to take it, and quickly. Only three days after Neale's death, on 26 December, Luttrell heard the news. "Dr. Newton, the mathematical professor, is advanced from being warden to master of the mint in the room of Mr. Neal, deceased; and sir John Stanley succeeds the Dr. as warden, a place worth 500l. per annum." Luttrell's use of the title "Dr." may provide the best clue to Newton's promotion. Recognized as England's leading intellectual, he was a figure in his own right who was able to command the post he wanted. Though Luttrell dated his news 26 December, the Mint accounts showed that Newton took office on 25 December, in which case the position was a birthday present. The letter patent confirming his appointment was finally sealed on 3 February 1700.

In assessing Newton's appointment as master, we need to keep two facts in mind. First, the progression from warden to master had no precedent at the Mint and was not repeated. Second, Newton still held both his fellowship and his chair in Cambridge. In fact, three-and-a-half years had sufficed to convert him into a civil servant. Far from wanting to return to Cambridge, he pursued the better position in order to ensure his continuation in London. He finally resigned both places in Cambridge in 1701, a year of heavy coinage when he took in nearly £3,500 as master, a sum that must have made the Cambridge income look too paltry to matter. Three-and-a-half years in London had been long enough also to teach him the facts of political life. Even with his patron out of office, he was able to secure the mastership he desired. And people worry about Catherine Barton's affair with Halifax!

From the Indenture, a formal contract between the master and the monarch, and from Newton's annual accounts, it is possible to determine his income as master. Over the period of twenty-seven years when he was master, his total profit from gold and silver coinage averaged £994 per annum. Beginning in 1703 and continuing at least through 1717 and probably longer, he received £150 per annum for managing the storage and sale of tin. For seven years, 1718–24, he earned an additional £100 in round numbers from the coinage of copper halfpennies and farthings. He received in addition an annual salary of £500. We know that he also received gifts, and he undoubtedly received other gifts of which we do not know. We cannot even estimate what their value might have been. According to statements he made in 1713, he had certain necessary expenses of about £180 per annum which he could not avoid. Newton's average income as master probably came to about £1,650. Individual years varied wildly, from £663 in 1703 (when his profit amounted to £13) to £4,250 in 1715 (when his profit swelled to £3,606). The average is misleading, because Newton's twenty-seven years as master included eleven years of the War of the Spanish Succession, which depressed the coinage. For the other sixteen years, his income as master averaged nearly £2,150, £2,250 during the years of copper coinage. To put his income in perspective, recall that the Treasurer's salary under Charles II was £8,000 per annum, though it was later reduced somewhat. No other official received half that. Most officials had various devices of dubious ethical status, such as Montague must have used in building his fortune, to increase their real incomes. The master of the Mint had less opportunity in similar enterprises. Newton would have disdained to engage in such in any case.

As I have mentioned, 1701 was a big year for the Mint. With a profit of £2,959, Newton's income almost reached £3,500. Already at the beginning of the year, nearly five years after his departure from Cambridge, he appointed William Whiston as his deputy in the Lucasian chair with the enjoyment of its full income. On 10 December, Newton formally resigned, enabling Whiston to become his successor, and about the same time he also resigned his fellowship at Trinity. He stood then eleventh in the order of seniority. He must soon have regretted his decision to surrender the income at Cambridge, for his profits at the Mint fell off quickly almost to nil. For five years, from 1703 through 1707, they did not get close to £100. Only with the peace did they recover to the level he must have expected when he pursued the position. At a time when £1,200 per annum for a single man was described as not merely easy but splendid, Newton's income as master was always considerable, and once the war ended, it ensured a life of material plenty, even under the more expensive conditions of London.

Studies of administrative history have pointed to the period around Newton's service in the Mint as the seed time of a professional civil service in England. They indicate the Treasury especially as the locus of such a development. The accounts from the Mint were not the only ones that began to arrive annually. In my opinion, Newton deserves to be recognized as a distinguished civil servant in the first age when such appeared. No doubt he did not transform the Mint. He did make it operate with more efficiency by far than it had shown in the past or would show again for another century.

ଈ

The office involved a further obligation, or potential obligation – membership in the House of Commons, where Newton could lend his support to the government, or perhaps to Halifax. As I have pointed out, Newton visited Cambridge in the election year 1698, though he did not stand. In 1701, he did. He was elected and served in the Parliament that began to sit on 20 December. As before in the Convention Parliament, he was not prominent in any respect. The death of William III led to a prorogation in May 1702, followed soon after by dissolution. Newton did not formally stand for the following election later that year. There had been unpleasantness at the former one. The defeated candidate, Anthony Hammond, had composed a pamphlet, *Considerations upon Corrupt Elections of Members to serve in Parliament;* although it contained no explicit reference to Cambridge, it did argue that the New East India Company was engaged in an extensive program of electoral corruption to ensure a governmental policy favorable to its interests. Halifax was associated with the New East India Company, having introduced the bill for its incorporation in 1698. Both Newton and the public at large could easily have read Hammond's pamphlet as a charge that he was a paid lackey. Furthermore, it hinted darkly that radical religious groups might subvert the Anglican church by the same device. In a letter to a friend (probably Bentley) in the summer of 1702, Newton indicated that he refused to come to Cambridge to stand openly in the new poll. Why would he not stand openly? Perhaps the role of religious conformity which Queen Anne herself injected into the election with her closing speech to the previous Parliament played some part in his decision. A pamphlet on the election of 1702 by the Jacobite James Drake, which referred specifically to Cambridge and to Halifax as a patron powerful in the election there, placed the issue of hypocrites, who were destroying the church by pretending to be true Protestants, at the center of attention. Drake's message was more unsettling than Hammond's. Newton instinctively ran for cover whenever similar arguments appeared.

Halifax, whose fortunes depended on a block of support in Commons, had a more aggressive attitude in mind, and he took care to prepare Newton's mind for the next contest. It came in 1705, and Halifax's desires prevailed. Newton fairly wore out the road to Cambridge with three visits. He was there on 16 April when the queen made a visit. At the final royal visit to the university that he would attend, Newton occupied a place on the stage. "The whole University lined both sides of the way from Emanuel College, where the Queen enter'd the Town, to the public Schools," Stukeley, who was an undergraduate then, recalled. "Her majesty dined at Trinity college where she knighted Sir Isaac, and afterward, went to Evening Service at King's college chapel . . . . The Provost made a speech to her Majesty, and presented her with a bible richly ornamented. Then she returned amid the repeated acclamations of the scholars and townsmen." The queen's "great Assistance" to Newton's election was his knighting, an honor bestowed not for his contributions to science, not for his service at the Mint, but for the greater glory of party politics in the election of 1705. Halifax, who had organized the visit, orchestrated it as a political rally. Besides Newton, the queen also knighted Halifax's brother and mandated that the university confer an honorary doctorate on Halifax himself. In a nonpartisan gesture, he allowed her also to knight Newton's old friend John Ellis, a mere academic then vice-chancellor of the university. Having gone back to London, Newton returned about 24 April and stayed on, soliciting votes, until the election on 17 May.

Things did not go well, and all of Newton's resolution could not prevent his being disturbed. Not only did he run dead last and not even close in a field of four, but the unpleasantness of 1701, which he had feared to face in 1702, repeated itself and took on a form which could not have upset him more. Simon Patrick, Bishop of Ely, described the scene to the House of Lords the following December as he urged an investigation into the corruption of youth – by Anglican zealots, of all things! – at the university: "at the election at Cambridge it was shameful to see a hundred or more young students, encouraged in hollowing like schoolboys and porters, and crying, No Fanatic, No Occasional Conformity, against two worthy gentlemen that stood candidates." Occasional conformity was the accepted practice by which dissenters qualified for full civil rights by taking the sacrament in the established church once a year. The move to revoke it, which was pushed by extreme Tories, struck at the heart of Newton's security. No scene could have shaken him more. No amount of encouragement from Halifax could tempt him to chance its repetition. The year 1705 marked the end of his career in Parliament.

Administrative duties continued at the Mint, however. What with a constant flow of references from the Treasury and memoranda in reply, the business of the Mint included much more than coining. It formed the pervasive background of Newton's life in London. With the advent of the War of the Spanish Succession in 1702, its demands did slacken, however. Beginning with 27 May 1703, the Mint did not coin a single day for nine months. It coined a total of seventy-five days during the ensuing six years. Under such circumstances, Newton was finally free, after seven years dominated by the administrative demands of his new institution, to consider other activities.

11

President
of the Royal Society

THE ROYAL SOCIETY, to which Newton had dedicated his *Principia* in 1687 only to ignore it steadfastly when he moved to London, stood at a low ebb during the early years of his residence in the capital city. Membership, which had reached more than two hundred in the early years of the 1670s, now scarcely numbered more than half that figure, and meetings, given over mostly to miscellaneous chitchat devoid of serious scientific interest, suggested little of the interests that had brought the society together forty years earlier. The presence of Robert Hooke, not Newton's favorite natural philosopher, may well have determined his absence from the weekly meetings. Hooke was usually there. When Newton put in one of his rare appearances to show a "new instrument contrived by him," a sextant, which would be useful in navigation, Hooke reminded him of past antipathies by claiming that he had invented it more than thirty years before. Hooke's death in March 1703 removed an obstacle and prepared the way for Newton's election as president at the next annual meeting on St. Andrew's Day, 30 November.

Obscurity covers the background to Newton's election. Spontaneous expressions of popular will did not govern the selection of officers of the Royal Society. In all probability, Dr. Hans Sloane, the secretary, made the prior arrangements. At the meeting on 30 November something nearly went awry. Newton was not a political leader who had only to be proposed to be elected. Only twenty-two of the thirty members present voted to place him on the council, a necessary preliminary to election as president. Once elected to the council, he still received only twenty-four votes for president. Clearly, a group within the Royal Society did not rush to welcome England's preeminent natural philosopher to the presidential chair.

Truth to tell, they did not rush to reelect him the next year, and the absence of vote totals in the society's Journal Book the following two years strongly implies a continuing want of universal enthusiasm.

Less than two years after Newton's election, Queen Anne knighted him in Cambridge. Master of the Mint and president of the Royal Society, Sir Isaac Newton had become a personage of consequence. The attention he devoted to his coat of arms testifies that he recognized as much. A year before his election, he had sat for another portrait by Kneller (Plate 3) and upon his election one by Charles Jervas. Now, as Sir Isaac Newton, he was painted by Sir James Thornhill (Plate 4) and by William Gandy. The rebel of yore had allowed the establishment wholly to co-opt him.

Newton did not attend the meeting of the society on 8 December, the first one after his election. On 15 December, he did appear and immediately took command. To the Royal Society he brought the same qualities he exercised at the Mint, administrative talent and a constitutional inability to slough off an obligation he had agreed to shoulder. In a history of the administration of the Royal Society, Sir Henry Lyons has stressed the paramount importance at this time in the society's affairs of vigorous and continuous leadership. After an interlude of absentee presidents chosen for their political prominence, the society watched with surprise as a man who had devoted his life to its announced goals seized its helm and bent his energy to steering it on a determined course. Newton made a point of running the council. It almost never met without him. Whereas Montague had attended one meeting of the council during the three years of his presidency and John Lord Somers none during his five, Newton failed to preside at a total of three meetings during the next twenty years before age began to restrict him. A society which had seen its president in the chair three times during the preceding eight years now saw him there for more than three meetings out of four. Newton frequently participated in the discussion at the meetings. His contribution to the Royal Society as its president was administrative rather than intellectual, however. It was not coincidence that the society's fortunes began to revive at much the time when he took charge of its affairs.

Administration did involve intellectual matters, of course. Newton was aware that meetings lacked serious content, and he came to the presidency armed with a "Scheme for establishing the Royal Society," intended to cure the disease. "Natural Philosophy," the "Scheme" declared, "consists in discovering the frame and operations of Nature, and reducing them, as far as may be, to general Rules or Laws, – establishing these rules by observations and experiments, and thence deducing the causes and effects of things . . . ." To this end, it might be convenient that one or two and ultimately perhaps three or four men skilled in the major branches of philosophy be established with pensions and bound to attend the weekly meetings. He went on to set down five major branches of natural philosophy, for each of which, presumably, he looked forward to appointing a pensioned demonstrator: mathematics and mechanics; astronomy and optics; zoology (to use our word), anatomy, and physiology; botany; and chemistry. He stated explicitly his intention that the society appoint only men who had established reputations in the sciences. In effect, Newton proposed the extension of an institution that began with the society itself, the curator of experiments, to provide solid substance at the weekly meetings. Newton's old nemesis Robert Hooke had filled that post with distinction for many years and by his efforts had kept the society afloat when the formless garrulity of the members had threatened to ground the meetings on utter banality. Newton did not mention Hooke, but his "Scheme for establishing the Royal Society" testifies adequately that he recognized what Hooke had done for the society. With Hooke dead, he set his first priority in replacing him – preferably in the plural.

He did more than hope. He found Hooke's replacement, Francis Hauksbee. We know nothing of Hauksbee's origin and background or of how Newton made his acquaintance. We know only that on 15 December 1703, at the first meeting over which Newton presided, Hauksbee appeared at the Royal Society for the first time and, though he was not then a member, performed an experiment with his newly improved air pump. He continued to attend the meetings; nearly every week, he showed an experiment using the air pump. In February, the council voted to pay him two guineas for his trouble, and in July, as it broke up for the summer, five guineas more. Hauksbee continued to serve the society, providing much of the scientific content of its meetings, for ten years until his death in 1713. Despite Newton's "Scheme," the society never did give him an official position. Each year the council voted him a gratuity, £15 for 1704–5, as much as £40 in some years, though the council occasionally reduced the sum when he had performed less energetically.

The precise nature of Hauksbee's relation to Newton cannot be defined with assurance. Left to himself, Hauksbee did not appear to have immense intellectual initiative. He devoted his first year and a half at the society to experiments with his air pump which showed little imagination and largely repeated experiments done earlier by Boyle and others. Later, he did go on to new themes, especially electricity and capillary action, in which his experiments exercised considerable influence on Newton. Of course, neither topic was previously unknown to Newton. No evidence tells us how far he may have guided Hauksbee's new ventures, though we must beware of attributing to Newton what was not his. There is no reason to think that he suggested mounting a globe of glass on an axle and thus nearly inventing the static-electric machine, though the effects that Hauksbee generated thereby powerfully stimulated Newton's imagination. The *Philosophical Transactions* published a steady stream of Hauksbee's experimental papers, and in 1709 he collected them together in his *Physico-Mechanical Experiments.* As a result, he became a famous scientist in his own right.

In 1707, Newton appeared for a time to have found a second demonstrator to supplement Hauksbee at meetings of the society. Dr. James Douglas frequently performed dissections at meetings, and in July of that year the council voted him a gratuity of £10. For reasons not recorded, the arrangement with Douglas failed to solidify. Although Douglas continued as an active participant in meetings, he never received another gratuity for his efforts.

There is no way to pretend that on the morrow of Newton's election the meetings of the Royal Society suddenly transformed themselves into profound discussions bubbling with philosophical ferment. The society's appetite for monstrosities brooked no satisfaction. Between dissections, Dr. Douglas showed them "a puppy Sufficiently Nourished whelpt alive about 10 Days ago which had no Mouth," and a week later he brought in its skull. In 1709, "There was shewed four Piggs all Growing to One Another taken out of a Sow after she was killed. M^r. Hunt was Ordered to give the Bearer three half Crowns and to preserve them in Spirits of Wine." Newton contributed his own share to the miscellaneous reflections, which threatened always to swamp serious scientific discussion, telling the members on one occasion of a man who died from drinking brandy and on another of a dog at Trinity that was killed by Macasar poison. At one meeting, he informed the society "that Bran Wetted and

Heated would bread Worms, which are Supposed to Proceed from Eggs Laid therein." Nevertheless the weekly meetings showed a steady improvement during Newton's presidency. From the low point in the 1690s, membership steadily grew and more than doubled during his years in office. No doubt many things contributed to the renewed vitality, but the enhanced level of the meetings, which Newton actively fostered, was not least among them.

❧

As he bent his energies to the countless miscellaneous minutiae of administration, Newton also reminded the Royal Society of its basic purpose in the most effective way possible. On 16 February 1704, from the chair, he presented it with his second great work, the *Opticks*. Unlike the printing of the *Principia*, we know nothing about the details of the *Opticks'* publication. Newton's election to the presidency may have figured in his decision finally to bring it out. John Wallis had been hectoring him about the book for nearly a decade. More recently, David Gregory had taken up the cry, and he recorded on 15 November 1702 that Newton "promised Mr. Robarts, Mr. Fatio, Capt. Hally & me to publish his Quadratures, his treatise of Light, & his treatise of the Curves of the 2^d Genre." He did not say when he would do so, however, and his elevation to the presidency may have provided the crucial stimulus. Newton did not dedicate the *Opticks* to the society, as he had the *Principia*. Nevertheless he did identify it with the society by letting the title page state that Samuel Smith and Benjamin Walford, printers to the Royal Society, were the publisher, and by mentioning the society in the "Advertisement" that functioned as preface.

Newton's election as president was not the only, or for that matter the primary, cause of the *Opticks'* publication in 1704, however. In the "Advertisement" he described briefly how he had composed most of it many years before. "To avoid being engaged in Disputes about these Matters, I have hitherto delayed the Printing, and should still have delayed it, had not the importunity of Friends prevailed upon me." Not many members of the Royal Society would have missed the veiled reference to Hooke, whose death in 1703 removed an obstacle both to Newton's presidency of the society and to his publication of the *Opticks*.

The "Advertisement" contained two more paragraphs, each of which alluded to further incentives to publish. One mentioned the "Crowns of Colours" that sometimes appear around the sun and the moon. The publication of Huygens's *Dioptrica* with his posthumous works in 1703 had included an account of such crowns. Newton apparently wanted to assert the independence of his own account of them. More significant was the third paragraph, which introduced the two mathematical papers "Tractatus de quadratura curvarum" ("A Treatise on the Quadrature of Curves") and "Enumeratio linearum tertii ordinis" ("Enumeration of Lines of the Third Order"), which Newton appended to the *Opticks*. Some years before, he stated, he had loaned out a manuscript with some general theorems about squaring curves, "and having since met with some Things copied out of it, I have on this Occasion made it publick . . . ." What he had met was a book published in 1703 by George Cheyne, *Fluxionum methodus inversa* (*The Inverse Method of Fluxions*). According to David Gregory, in a memorandum of 1 March 1704, "Mr. Newton was provoked by Dr. Cheyns book to publish his Quadratures, and with it, his Light & Colours, &c." The issue was rather more complicated than Gregory's note implied. In his collection of anecdotes, Conduitt included a story he heard from Peter Henlyn that when Cheyne came to London from Scotland, Dr. Arbuthnot introduced him to Newton and told him about the book Cheyne had written but could not afford to publish. Cheyne also reported later that he had shown the manuscript to Newton, who "thought it not intolerable." As Conduitt heard it, Newton offered Cheyne a bag of money; Cheyne refused to take it. Both were thrown into confusion by the impasse, and Newton refused to see him any more. Cheyne must have been startled to read the extent of Newton's resentment of his assertion of independence in the "Advertisement." It possibly explains why Cheyne quickly dropped out of the Royal Society and chose to pursue his career in medicine rather than in mathematics and natural philosophy.

As far as the mathematical papers were concerned, Newton merely published expositions, composed a decade earlier, which summed up work that stretched back nearly forty years. Moreover, because Leibniz and his followers had been publishing a method identical to the basic concepts in "De quadratura" for a number of years, the appearance of the short treatise in 1704 did not constitute a major event in the history of mathematics. The papers marked an epoch for Newton nevertheless. At long last, after more than thirty years of delay and evasion, he published some mathematical work. If it was too late now to head off the battle with Leibniz, at least he could show the world something of the substance behind his reputation as a mathematician.

The situation in regard to the *Opticks* partially repeated that in regard to the mathematical papers. As far as our understanding of Newton's scientific thought is concerned, the *Opticks* contained nothing new. With only the smallest exceptions, it presented work he had completed more than thirty years earlier, and the exceptions belonged to the early 1680s. Unlike the fluxional method, however, the *Opticks* had not been duplicated by another investigator. In 1704, only a few men had digested the import of Newton's published paper of 1672. Hence the impact of the *Opticks* virtually equaled that of the *Principia*. Indeed, it may have exceeded it, for the *Opticks*, written in prose rather than in geometry, was accessible to a wide audience as the *Principia* was not. Through the eighteenth century, it dominated the science of optics with almost tyrannical authority and exercised a broader influence over natural science than the *Principia* did. More than one of Newton's younger contemporaries, including his disciple John Machin, told Conduitt that there was more philosophy in the *Opticks* than in the *Principia*. The work remains permanently one of the two pillars of Newton's imperishable reputation in science.

The *Opticks* that Newton published in 1704 was not the *Opticks* he had planned in the early 1690s. That work had reached its climax in a Book IV dedicated to the demonstration that forces that act at a distance exist. As he had done on other occasions, he finally now shrank from laying so much bare in public. In the *Opticks* he did publish, he eliminated Book IV and focused the work sharply on optical problems – the theory of colors and the allied concept of the heterogeneity of light. There is no need to repeat his demonstration of the two. Suffice it to say, they constitute an enduring legacy to the science of optics. Somewhat unobtrusively, Newton did slip in assertions that optics requires forces acting at a distance similar to the *Principia*'s force of gravity. In Book II, he pointed out that reflection cannot be caused by light impinging on the solid parts of bodies. The reflection of light from the back side of a glass in a vacuum argues against the impact theory of reflection; the uniformity of reflection from a surface, which would require the perfect alignment of all its particles, argues the case more strongly.

And this Problem is scarce otherwise to be solved, than by saying that the Reflexion of a Ray is effected, not by a single point of the reflecting Body, but by some power of the Body which is evenly diffused all over its Surface, and by which it acts upon the Ray without immediate Contact. For that the parts of Bodies do act upon Light at a distance shall be shewn hereafter.

He proceeded then to argue that bodies reflect and refract light by one and the same power, and from a table that compared refractive power to density, both for bodies in general and for the special class of "fat sulphureous unctuous Bodies," he concluded that the reflecting and refracting power arises from the sulfurous parts bodies contain. The argument assumed the corpuscular conception of light, of course.

Book III, Newton's still brief investigation of diffraction, contained the promised demonstration that a body "acts upon the Rays of Light at a good distance in their passing by it." In a set of sixteen Queries, the first incarnation of the famous Queries that conclude the *Opticks* and Newton's substitute in the published work for the suppressed Book IV, he went on to consider forces further in an explicitly speculative context. The Queries of 1704–6 were Newton's last major publication of hitherto unknown scientific work, the culminating assertion of the Newtonian program in natural philosophy before the timidity of age and his progressive domestication as he settled ever more comfortably into the seat of authority and power led the quondam rebel to compromise some of his bolder positions. We read the Queries today as they were published in the third English edition, really as they were published a few years earlier in the second English edition, as he altered little after it. Among the final thirty-one, Queries 17–24 assert the existence of a universal aether and offer an explanation of forces in terms of such. To understand the original set of Queries it is necessary to remind oneself that they ended with number 16 and contained no suggestion of an aether to modify the assertions they made under the guise of rhetorical questions. Not all of the Queries concerned forces; the first ones did.

Query 1. Do not Bodies act upon Light at a distance, and by their action bend its Rays; and is not this action (*caeteris paribus*) strongest at the least distance?

. . .

Qu. 4. Do not the Rays of Light which fall upon Bodies, and are reflected or refracted, begin to bend before they arrive at the Bodies; and are they not reflected, refracted, and inflected [diffracted], by one and the same Principle, acting variously in various Circumstances?

Qu. 5. Do not Bodies and Light act mutually upon one another; that is to say, Bodies upon Light in emitting, reflecting, refracting and inflecting it, and Light upon Bodies for heating them, and putting their parts into a vibrating motion wherein heat consists?

. . .

Qu. 7. Is not the strength and vigor of the action between Light and sulphureous Bodies observed above, one reason why sulphureous Bodies take fire more readily, and burn more vehemently than other Bodies do?

Though he couched them as questions, no one is apt to mistake the positive answers Newton intended. It was a less explicit form than he once had planned; nevertheless the *Opticks* did set forth the Newtonian program in natural philosophy.

"My Design in this Book," Newton began the *Opticks*, "is not to explain the Properties of Light by Hypotheses, but to propose and prove them by Reason and Experiments." The statement was all that remained of a projected introduction in which he flung down a methodological gauntlet to mechanical philosophers to match the metaphysical one. There is a twofold method (he argued in the suppressed introduction), of resolution and composition, which applies to natural philosophy as well as to mathematics, "& he that expects success must resolve before he compounds. For the explication of Phaenomena are Problems much harder then those in Mathematicks." He described the method in terms nearly identical to those he later used in Query 31, with echoes also of a passage later embodied in his third rule of reasoning in philosophy.

Could all the phaenomena of nature be deduced from only thre or four general suppositions there might be great reason to allow those suppositions to be true: but if for explaining every new Phaenomenon you make a new Hypothesis if you suppose y^t ye particles of Air are of such a figure size & frame, those of water of such another, those of Vinegre of such another, those of sea salt of such another, those of nitre of such another, those of Vitriol of such another, those of Quicksilver of such another, those of flame of such another, those of Magnetick effluvia of such another, If you suppose that light consists in such a motion pression or force & that its various colours are made by such & such variations of the motion & so of other things: your Philosophy will be nothing else then a systeme of

Hypotheses. And what certainty can there be in a Philosophy wch consists in as many Hypotheses as there are Phaenomena to be explained. To explain all nature is too difficult a task for any one man or even for any one age. Tis much better to do a little with certainty & leave the rest for others that come after, than to explain all things by conjecture without making sure of any thing.

Even as president of the Royal Society, Newton did not find it easy to express fundamental convictions in public. He feared criticism. He preferred silence to the risk of controversy in which he might find himself made an object of ridicule. It tells us much about his unassuageable sense of insecurity that even from the pinnacle of renown he occupied in 1704 he suppressed the polemical introduction and did not dare to publish the once projected Book IV, though the suggestions he did publish argue that it expressed his beliefs. The Queries that he did allow to pass were brief, filling only two-and-a-half manuscript pages at first. Nevertheless, they represented a considerable step for Newton, who until this time had not allowed any more than hints about his convictions on the ultimate nature of things to appear in print. And having taken one step he found it possible to take others. Perhaps I should say rather that he found it impossible not to, for the Queries acquired an independent existence and took command over Newton as other things had done in their turn. Apparently his original intention had been a set of short, staccato questions of one or two sentences, such as Queries 1–7 permanently remained. By the time he reached Query 12, however, he felt impelled to write a bit more, about a third of a sheet in all, and so also for Queries 13 and 15. Queries 10 and 11 no longer seemed adequate, and he returned to expand them. Here he did manage to stop for the first edition, but the preparation of that edition nearly merged with the preparation of the Latin edition, which followed two years later. In the Latin edition, he further expanded Query 10, and, more important, he added seven new Queries all but one of which were longer than the original sixteen taken together. In the new Queries, Newton expressed fundamental views on the nature of light, on the nature of bodies, on the relation of God to the physical universe, and on the presence in nature of a whole range of forces which furnish the activity necessary for the operation of the world and for its permanence. At the last moment he dared even a bit more and inserted three further speculative passages in the Addenda to the volume.

The new Queries were the most informative of the speculations that

Newton ever published. In the second English edition, he added eight more Queries, which he inserted as numbers 17–24 between the first and second sets. Hence the seven Queries added to the Latin edition appear in all later editions, including those in general circulation today, as Queries 25–31. To avoid confusion, I shall refer to them by their final numbers, though they carried different ones when they first appeared. A point important for understanding the Queries new to the Latin edition, especially Queries 29 and 31, lies behind the difference in number. The third and final set of Queries, 17–24, are the ones that assert the existence of an aether that pervades all space. When he added them to the second English edition, Newton also put in passages about a second subtle fluid found in the pores of bodies that causes electrical phenomena (static electrical, of course) when it is agitated. In his final years, a growing philosophic caution led Newton to retreat somewhat toward more conventional mechanistic views, although his subtle aethers composed of particles repelling one another always remained more sophisticated than the clumsy fluids of standard mechanical philosophies. When he originally published Query 31, no suggestion of an aether or fluid modified its rhetorical questions about the prevalence of forces between bodies at every level of phenomena.

Query 31 was an extended version of the speculations on forces that Newton had once planned to insert in the *Principia.* Heavily – indeed, overwhelmingly – chemical in content, it was arguably the most advanced product of seventeenth-century chemistry. What had Newton drawn from chemistry? The conviction that its phenomena require the presence of forces between particles for their explanation.

Have not the small Particles of Bodies certain Powers, Virtues, or Forces, by which they act at a distance, not only upon the Rays of Light for reflecting, refracting, and inflecting them, but also upon one another for producing a great Part of the Phaenomena of Nature? For it's well known, that Bodies act one upon another by the Attractions of Gravity, Magnetism, and Electricity; and these Instances shew the Tenor and Course of Nature, and make it not improbable but that there may be more attractive Powers than these. For Nature is very constant and comfortable to her self.

The body of the Query detailed the evidence, both from chemistry and elsewhere, on which his reasoning rested. The remarkable continuity in Newton's lifelong investigation of the nature of things revealed itself in

the appearance of all the crucial phenomena that had captured his attention more than forty years earlier when, as an undergraduate, he composed an earlier set of "Quaestiones." Together with the chemical phenomena, they now seemed to require the admission of forces of attraction
and repulsion between particles. Thus nature will be very comfortable to
herself, he concluded, performing all her great motions by the attraction
of gravity and her small ones by the forces between the particles. Moreover, nature requires the presence of active principles. Inertia, the basic
concept of conventional mechanical philosophies, is a passive principle by
which bodies persevere in their motions. Phenomena reveal, however,
that nature contains sources of activity, active principles which can generate new motions. Still worried by the hostile reception of his dynamic
conception of nature, Newton felt obliged to add a disclaimer.

These Principles I consider, not as occult Qualities, supposed to result from the
specifick Forms of Things, but as general Laws of Nature, by which the Things
themselves are form'd; their Truth appearing to us by Phaenomena, though their
Causes be not yet discover'd. For these are manifest Qualities, and their Causes
only are occult. . . . To tell us that every Species of Things is endow'd with an
occult specifick Quality by which it acts and produces manifest Effects, is to tell us
nothing: But to derive two or three general Principles of Motion from Phaenomena, and afterwards to tell us how the Properties and Actions of all corporeal
Things follow from those manifest Principles, would be a very great step in
Philosophy, though the Causes of those Principles were not yet discover'd: And
therefore I scruple not to propose the Principles of Motion above-mention'd, they
being of very general Extent, and leave their Causes to be found out.

In the first paragraph on Query 31, Newton inserted a caution about
his assertion of the existence of forces. "How these Attractions may be
perform'd, I do not here consider. What I call Attraction may be perform'd by impulse, or by some other means unknown to me. I use that
Word here to signify only in general any Force by which Bodies tend
towards one another, whatsoever be the Cause." We read the passage
today after the Queries on the aether and after a paragraph, inserted in
the second English edition to conclude Query 29, which refers the meaning of the word *attraction* to those Queries. In the Latin edition of 1706,
the reservation recalled rather the conclusion of Query 28. In Query 28,
the refutation of wave theories of light led Newton into an argument
against the possibility of a dense, Cartesian aether's filling the heavens,

and thence into an explication of his ultimate objection against conventional mechanical philosophies, their tendency to make nature self-sufficient and thus to dispense with God. Some ancient philosophers, he argued, took atoms, the void, and the gravity of atoms as the first principles of their philosophy and attributed gravity to some other cause than matter.

Later Philosophers banish the Consideration of such a Cause out of natural Philosophy, feigning Hypotheses for explaining all things mechanically, and referring other Causes to Metaphysicks: Whereas the main Business of natural Philosophy is to argue from Phaenomena without feigning Hypotheses, and to deduce Causes from Effects, till we come to the very first Cause, which certainly is not mechanical; and not only to unfold the Mechanism of the World, but chiefly to resolve these and such like Questions. What is there in places empty of Matter, and whence is it that the Sun and Planets gravitate towards one another, without dense Matter between them? Whence is it that Nature doth nothing in vain; and whence arises all that Order and Beauty which we see in the World? . . . How do the Motions of the Body follow from the Will, and whence is the Instinct in Animals? Is not infinite Space the Sensorium of a Being [Annon Spatium Universum, Sensorium est Entis] incorporeal, living, and intelligent, who sees the things themselves intimately, and thoroughly perceives them, and comprehends them wholly by their immediate presence to himself . . . .

David Gregory, who held an extensive discussion of the new Queries with Newton on 21 December 1705, recorded the interpretation of this passage in a memorandum.

His Doubt was whether he should put the last Quaere thus. *What the space that is empty of body is filled with.* The plain truth is, that he believes God to be omnipresent in the literal sense; And that as we are sensible of Objects when their Images are brought home within the brain, so God must be sensible of every thing, being intimately present with every thing: for he supposes that as God is present in space where there is no body, he is present in space where a body is also present. But if this way of proposing this his notion be too bold, he thinks of doing it thus. *What Cause did the Ancients assign of Gravity.* He believes that they reckoned God the Cause of it, nothing els, that is no body being the cause; since every body is heavy.

At the last moment – after the last moment, really – Newton decided that he had indeed been too bold. He tried to recall the whole edition; and from all the copies he could lay his hands on, he cut out the relevant page and pasted in a new one which asserted, not that infinite space is the sensorium of God, but that "there is a Being incorporeal, living, intelligent, omnipresent, who in infinite Space, as it were in his Sensory, [tanquam Sensorio suo] sees the things themselves intimately . . . ." He failed to alter every copy, however, and one of the originals made its way to Leibniz, who did not fail to hold up to ridicule the concept of space as the sensorium of God. In its initial form the passage recalled Newton's earlier paper "De gravitatione," the beginning of his rebellion against Cartesian philosophy because of its atheistical tendencies. Following the implications of the rebellion, he had traveled far. In the Latin edition of the *Opticks*, he gave the fullest exposition of his own conception of nature he would ever put in print before, in his old age, he tried to placate critics by seeming retreats to more conventional positions.

☙

Meanwhile, his presidency of the Royal Society had led Newton to an unhappy resumption of relations with another former acquaintance, John Flamsteed. Only a few months after his election, on 12 April 1704, Newton went down to Greenwich, enquired about the state of Flamsteed's observations, and when he was shown them asked to recommend them to Prince George, Queen Anne's consort, for his financial support of their publication. In the light of Newton's later actions, there is only one reasonable interpretation of the visit. Still tormented by his failure with lunar theory and still convinced that Flamsteed had caused it, he determined to exercise his authority as president of the Royal Society to get Flamsteed's observations in order, as he put it, to have another stroke at the moon. A perfected lunar theory would crown a second edition of the *Principia*. Thus Newton was all benevolent philanthropy on the visit in April despite Flamsteed's suspicions. "Do all the good in your power," he told Flamsteed as he left, and Flamsteed, typically, noted in his recollections that such had always been the rule of his life, "though I do not know that it ever has been of his." Newton's philanthropy had given way to despotic ire long before Flamsteed made his sour comment.

In the autumn, as preliminary negotiations proceeded, Flamsteed took an action that passed the initiative from his hands to Newton's. Early in November, he drew up an "Estimate" of what the projected *Historia britannica coelestis* (*British History of the Heavens*) would contain, in effect an objective account of his achievement at Greenwich, and he gave it to

James Hodgson, a former assistant who had married his niece, to show around the Royal Society as evidence of his accomplishment. Newton, who was in the chair, could not resist the opportunity thus presented to seize control and ensure his access to the precious observations. By the annual meeting two weeks later, the society had approached the prince with Flamsteed's estimate of the number of pages involved, and the prince had expressed his interest in very positive terms. To facilitate matters, the society proceeded at once to elect Prince George to membership, and before December was out it received a letter from his secretary in which "the President was desired to take what Care in this Matter he shall think Necessary Towards the most Speedy publication of so usefull a Work . . . ." More than ten years passed, years filled with inexpressible bitterness, before Flamsteed succeeded on the eve of his death in pushing Newton back out of his affairs.

Flamsteed cannot escape his own share of responsibility for the debacle that followed. Whatever his faults, however, Newton was the primary cause. Though Flamsteed was a member of the Royal Society, it never occurred to Newton to include him in the delegation that waited on Prince George about his work. Worse, the referees appointed to survey Flamsteed's papers and make recommendations, referees of whom Newton was the leader of course, systematically ignored Flamsteed's carefully thought-out plan of publication. The plan was neither arbitrary nor silly. Eventually, after Flamsteed's death two of his devoted assistants did complete the *Historia coelestis* according to his plan, and it is recognized by qualified experts as one of the great landmarks of the science of astronomy. Flamsteed wanted to set his catalogue squarely in its historical tradition by publishing in the same volume all the significant prior catalogues from Ptolemy to Hevelius. Knowing that the catalogue he wanted to give to the world, the monument of a lifetime's devoted labor, was not yet done, he asked for money to hire calculators to complete the reduction of his observations. The referees said not a word about the earlier catalogues. They recommended £180 for calculators to compute "the places of ye moon and planets & comets" – that is, the information Newton wanted. Implicitly they treated the catalogue of stars as complete in its present state by allowing nothing for further computations on it; and Newton's demand to proceed with it at once became a bone of contention. One would think Newton might have considered that the man who had given his life to making the observations – which the referees told the prince were "the fullest & Complatest" ever made, so that their loss would be irreparable – could be trusted to present them properly. Quite the contrary; convinced that he alone fully understood, Newton pressed forward and succeeded in needlessly depriving the world of the observations and the catalogue for another twenty years.

After they had reviewed Flamsteed's papers, the referees appointed by the Royal Society pursuant to the instructions in the letter from Prince George's secretary – Newton, Wren, Gregory, Francis Robartes, and Dr. John Arbuthnot – proceeded with arrangements. Flamsteed was galled that the bookseller (or publisher) Awnsham Churchill was to receive a profit, whereas the referees were unwilling even to consider what Flamsteed called "an honorable recompense for my paines, and 2000*lib*. in expense." Newton quickly seized the importance of the recompense to Flamsteed and for no apparent reason other than spite refused to hear of it. The issue acquired a further dimension. As soon as the prince accepted the budget with an item for calculators, Flamsteed hired two and set them to work – on the fixed stars, to be sure, not on planets, comets, and the moon as the referees intended. In a short time he ran up a bill of £173; Newton kept him waiting three years before he released £125 to him.

Negotiations on the articles of agreement used up nearly all of 1705. Among the issues on which Flamsteed had to give way was the decision to print the catalogue in volume 1. That is, the articles of agreement effectively specified that the catalogue of fixed stars to be published would be the currently existing catalogue, not a catalogue yet to be completed. As the articles demanded, Flamsteed immediately gave Newton the manuscript copy for volume 1, the catalogue excepted. Newton would not allow work to begin until he had a copy of the catalogue in his hands. After some backing and filling, Flamsteed agreed early in March to give Newton a copy of the catalogue as it then stood. He insisted that the copy be sealed. Later, Flamsteed loudly protested Newton's perfidy in breaking his word and opening the catalogue. There is no evidence whatever that Newton ever accepted Flamsteed's condition, however. From the time he received it, Newton knew what the catalogue lacked. Although he could have received that information from Flamsteed by word of mouth, he probably treated the catalogue as an open manuscript from the beginning. There is no point in pursuing the issue of the sealed catalogue at great length. Too much attention has focused on it. Flamsteed had grievances aplenty. It appears to me that he grasped at the alleged breaking of the seal to give specific content to his wholly justified sense of outrage at the way he was treated.

Finally, on 16 May 1706, the first sheet was printed off. Flamsteed could not contain his excitement. On 19 May, he went to Newton's house to collect his observation notes for use in correcting the proofs. Newton told him they must go slowly at first. By 24 May, when he had not yet received the second sheet, he wrote Churchill in admonishment. He wrote Churchill again on 6 June, strongly dissatisfied both with the rate of progress and with the accuracy. On 7 June, a Friday, he went to Churchill's office; though the printer did not come when summoned, he did send the fourth sheet, signature D. Flamsteed returned it corrected Monday morning, 10 June. And so it continued, with Flamsteed in hot haste to forward the press, while Churchill, despite the agreement to print five sheets per week, barely produced one.

Early in 1708, with the completion of the manuscript Flamsteed had delivered, the question of the catalogue could no longer be evaded. Would it go in volume 1 as Newton wanted or in volume 3 as Flamsteed wished? The press came to a halt. Finally, on 20 March 1708, all parties concerned met at the Castle Tavern and agreed that Flamsteed would hand over his observations with his most advanced instrument, the mural arc, and another catalogue of the fixed stars that he brought to the meeting, that he would further correct the deficiencies of the catalogue he had given two years before, that Newton would issue £125 to him, and that upon delivery of the catalogue of the fixed stars "as far as it can be Compleated at this Time" he would receive the rest of the money due him. He did finally receive £125 in April. In fact nothing more had been done by October, when Prince George died and the project of necessity came to a halt. Newton vented his frustration by having Flamsteed's name stricken from the list of the Royal Society in 1709 for nonpayment of dues – though he ignored plenty of others no less in arrears. Flamsteed used the respite to do what he wanted to do anyhow; he finished his catalogue.

&

In the years following 1710, Newton steadily consolidated his position within the society. At the end of 1713, Sloane decided to step down after twenty years as secretary. Apparently he made the decision under some pressure. Halley, a man clearly identified as a Newtonian, who would not be an independent factor in the society, succeeded Sloane. Scarcely a year after Sloane's resignation, the other secretary, Richard Waller, died in January 1715. Another identifiable Newtonian, Brook Taylor, succeeded him. Taylor resigned at the end of 1718 to be succeeded by John Machin, whom Newton's support had earlier helped to place in the Gresham chair of astronomy. Three years later, James Jurin, a protégé of Bentley from Trinity, replaced Halley. Beyond the officers, the society in these years acquired an increasing number of young, active members, natural philosophers and mathematicians, who can be described only as Newtonians – John Craig, William Jones, John and James Keill, John Freind, Roger Cotes, Robert Smith, Colin Maclaurin, J. T. Desaguliers, Henry Pemberton – and if "Newtonian" is not a meaningful description of physicians, there were nevertheless two who identified themselves with him, Richard Mead and William Cheselden.

Several passages in William Stukeley's *Memoirs* suggest the extent of Newton's domination of the society during his final years. The council was composed, Stukeley said, of the older members who had been of some service and who were chosen by rotation to acquaint them with the society's administration.

He regarded the choice of useful members, more than the number, so that it was a real honor. Nor did any presume to ask it, without a genuine recommendation, and having given some proofs of their abilities. They were then previously to be approved of by the Council, where their qualifications were freely canvased; therefore less lyable to be balloted for with partiality or prejudice.

Elsewhere, Stukeley made it clear just who canvassed qualifications.

In November 1725 I was again auditor of the accounts of the Royal Society: we din'd with Sir Isaac, and after dinner we desired him to recommend the Council to be elected on S. Andrews day approaching: which he did. I have now the paper of his own writing, . . . the names of the Council for the ensuing year, among which he put down mine.

Stukeley also mentioned that in 1721 when Halley retired, a number of members, including Hans Sloane and Lord (formerly Sir John) Percival, induced him to stand for secretary against Newton's opposition. Stukeley lost, though by a small majority. "Sir Isaac show'd a coolness toward me for 2 or 3 years, but as I did not alter in my carriage and respect toward him, after that, he began to be friendly to me again."

An almost imperial tone crept into the society following 1710. At the council meeting on 20 January 1711, four proposals "were thought fit to be made Orders of the Council" and were read at the next meeting of the society. They included the following:

1. That no Body Sit at the Table but the President at the head and the two Secretaries towards the lower end one on the one Side and the other Except Some very Honorable Stranger, at the discretion of the President.

. . .

3. That no person or persons talk to one-another at the Meetings, or So loud as to interrupt the business of the Society but address themselves to the President

At some point, Newton also established a practice that the mace be placed on the table only when the president was in the chair. Sloane's first act after his election following Newton's death was to decree that the mace appear on the table at every meeting regardless of who was presiding.

In the years following 1710, the level of the meetings continued to rise. Hauksbee usually had an experiment, on electricity or capillary action or the refraction of light. He died in 1713, but early in 1714 Newton discovered his replacement, J. T. Desaguliers. Really he found two replacements, for in William Cheselden he appeared to have the anatomical demonstrator he had sought in vain. In the summer of 1714, the council voted to excuse both from the weekly payments in view of their expected usefulness to the society. Cheselden never worked out, probably because his surgical practice thrived too well. Now and then, rather infrequently, he performed demonstrations at meetings, but the society under Newton never did acquire the anatomical demonstrator he wanted. Desaguliers, however, became a fixture at the meetings, where he carried out sets of experiments intimately related to various aspects of Newtonian natural philosophy. Some of his experiments, such as the transmission of heat through a vacuum, influenced Newton's views, and others found their way into the third edition of the *Principia*. With Desaguliers as the mainstay and other young Newtonians such as Jurin, Taylor, and Keill frequently contributing papers, the meetings picked up again for a few years. In an amateur society their vitality was always tenuous, however, and in Newton's old age, when a firm hand no longer grasped the helm, they slackened once more.

❧

Amid Newton's various successes in the Royal Society, one failure continued to nag him, the edition of Flamsteed's observations. With the death of Prince George in 1708, the authority of the referees had lapsed, and the project had come to a halt. Nothing happened for two years except that Flamsteed seized the respite finally to complete his catalogue of fixed stars to his own satisfaction. As far as we know, the two men did not communicate. From Flamsteed's point of view, Newton had used his power to obstruct the publication of a work which would detract from his renown. Newton saw the episode with the colors reversed, and in the end he could not tolerate the thought that Flamsteed had denied him the observations he needed. Indeed he now needed them even more, for he had finally engaged himself to publish a second edition of the *Principia*. On 14 December 1710, before a special meeting of the council, Dr. Arbuthnot, one of the former referees, who was also physician to Queen Anne, suddenly produced a warrant by which the queen appointed the president of the society and such others as the council thought fit to be "constant Visitors" of the Royal Observatory. The word *visitor* in this usage derived from ecclesiastical sources and referred to one authorized to visit an institution officially for the purpose of inspection and supervision in order to prevent or remove abuses or irregularities. Visitors could be appointed either for specific occasions or for continuing supervision. Hence the adjective *constant* placed the Observatory permanently under the control of the Royal Society. We know nothing of the background of the warrant. Flamsteed never doubted that Newton had engineered it to put him and the Observatory at the president's mercy. Newton consistently used it to that end, and it stretches credulity to question Flamsteed's account.

On the virtual morrow of the warrant, on 14 March 1711, Dr. Arbuthnot wrote to Flamsteed that the queen had "commanded" him to complete the publication of the *Historia coelestis*. He asked Flamsteed to deliver the material still lacking, primarily the catalogue of the fixed stars. Flamsteed did not respond immediately. He must have inquired discreetly what was happening. According to his letter to Abraham Sharp in the middle of May, he learned on 25 March that the printing of the catalogue had already commenced. That same day, he finally sent to Arbuthnot a reply he had composed earlier, welcoming the news that publication had resumed and informing him that he had completed the catalogue as far as he thought necessary. Meanwhile, "ye good providence of God (that has hitherto conducted all my Labours; and I doubt not will do so to an happy conclusion)" had led him to further fresh discoveries. Finding how far the planets' observed places differed from existing tables, he had begun to construct new ones. He needed help to complete them and make the work worthy of the queen's patronage and the memory of her consort. He asked Arbuthnot to meet him to discuss the matter. The mask now dropped, and Flamsteed received a furious letter, not from Arbuthnot but from Newton, whose patience did not survive even the first stage of the renewed trial.

S^r

By discoursing wth Dr Arbothnot about your book of observations wch is in the Press, I understand that he has wrote to you by her Ma^{ts} order for such observations as are requisite to complete the catalogue of the fixed stars & you have given an indirect & delatory answer. You know that the Prince had appointed five gentlemen to examin what was fit to be printed at his Highness expence, & to take care that the same should be printed. Their order was only to print what they judged proper for the Princes honour & you undertook under your hand & seal to supply them therewith, & thereupon your observations were put into the press. The observatory was founded to the intent that a complete catalogue of the fixt stars should be composed by observations to be made at Greenwich & the duty of your place is to furnish the observations. But you have delivered an imperfect catalogue wthout so much as sending the observations of the stars that are wanting, & I heare that the Press now stops for want of them. You are therefore desired either to send the rest of your catalogue to Dr Arbuthnot or at least to send him the observations wch are wanting to complete it, that the press may proceed. And if instead thereof you propose any thing else or make any excuses or unnecessary delays it will be taken for an indirect refusal to comply wth her Maj^{ts} order. Your speedy & direct answer & compliance is expected.

Although Flamsteed was certainly stalling for time, his letters to Sharp during 1710 demonstrate that the new planetary tables were not a pretense.

Early in April, Flamsteed confirmed the rumor that the catalogue was already being printed by getting a completed sheet into his hands, although Arbuthnot had explicitly assured him that the rumor was not true. He confronted Arbuthnot with his lie about the catalogue and with alterations that Halley, his enemy, had made. Arbuthnot evaded the lie, told Flamsteed the alterations had been made to please him, argued that they did not matter, and concluded by bluntly informing him that they would calculate the rest of the catalogue from his observations if he did not send it. More than the catalogue was involved. Though it had not come out in earlier correspondence, Flamsteed knew that Newton did not intend to publish all of his observations with the mural arc, observations which Flamsteed regarded as the empirical foundation of the catalogue and the voucher to its accuracy. After thirty-five years of labor, he faced the brutal fact that his enemies held the power to publish his life's work in a form which, in his eyes, mutilated and spoiled it. Confronted with *force majeure,* Flamsteed gathered his righteousness about himself and for once transformed it into dignity and indeed heroism.

I have now spent 35 years in composeing & Work of my Catalogue which may in time be published for ye use of her Majesty's subjects and Ingenious men all ye world over [he wrote to Arbuthnot]: I have endured long and painfull distempers by my night watches & Day Labours, I have spent a large sum of Money above my appointment, out of my own Estate to compleate my Catalogue and finish my Astronomical works under my hands: do not tease me with banter by telling me y^t these alterations are made to please me when you are sensible nothing can be more displeasing nor injurious, then to be told so.

Make my case your own, & tell me Ingeniously, & sincerely were you in my circumstances, and had been at all my labour charge & trouble, would you like to have your Labours surreptitiously forced out of your hands, convey'd into the hands of your de[c]lared profligate Enemys, printed without your consent, and spoyled as mine are in ye impression? would you suffer your Enemyes to make themselves Judges, of what they really understand not? would you not withdraw your Copy out of their hands, trust no more in theirs and Publish your own Works rather at your own expence, then see them spoyled and your self Laught at for suffering it. . . . I shall print it alone, at my own Charge, on better paper & with fairer types, then those your present printer uses; for I cannot bear to see my own Labours thus spoyled . . . .

Withdraw his copy he could not. Newton had it. Withdraw cooperation he could, though only at the risk that his life's work might never see the light of day. The biographer of Newton unhappily searches in vain through the whole dismal chronicle of closet tyranny for an action one-tenth as creditable.

The die was cast. Newton, with Halley following his direction, proceeded directly to complete the publication of the *Historia coelestis* early in 1712. A large folio volume, it began as Newton had always intended it should begin, with the catalogue of fixed stars, the copy Flamsteed handed to Newton in 1708 with the missing constellations filled in by 500 stars

that Halley calculated from Flamsteed's observations. Thus the catalogue, like the one Flamsteed planned, increased the number of stars
charted by astronomy from approximately one thousand to approximately
three thousand. The volume went on to the observations made before
1689, which had been printed in 1706–7, and then to observations of the
planets, the sun and moon, and eclipses of the satellites of Jupiter made
with the mural arc. Nowhere did Newton's determination to bend
Flamsteed's work to his own purposes appear more clearly. He eliminated
the observations of the fixed stars, the results of which appeared of course
in the catalogue, and printed those of use to him. A shameful preface
asserted that Flamsteed had not wanted to give his observations out and
that only the order of Prince George and the industry of the referees had
secured their publication. Flamsteed had planned a long Prolegomena to
describe his methods of observation to justify the degree of accuracy the
catalogue claimed: units of five seconds, an increase in accuracy of one
order over previous catalogues. Not only did the preface say nothing to
this effect, but it also cast doubt on the whole enterprise by suggesting
that Halley had had to correct various deficiencies. Newton arranged for
the government to pay Halley £150 for his work, £25 more than
Flamsteed would have received for more work. Newton was then revising
the *Principia* for the second edition. As he had formerly done with Hooke,
he went through the first edition systematically removing every reference
to Flamsteed that he could. Because his successful attack on the comet of
1680–1 rested primarily on Flamsteed's observations, for which he had
no substitute, he was not quite able to reduce him to nonexistence. He did
cancel his name in fifteen places.

   Nor was Newton done yet. He intended to bring Flamsteed to heel. As
Visitor of the Observatory, he summoned him to a meeting at Crane
Court, the home of the Royal Society, on 26 October 1711, with Sloane,
Mead, and himself, to inform them if the instruments were in order and
fit to carry on observations. The meeting was a mistake. Of all the men
alive, Flamsteed knew best how to drive Newton wild. He was never in
better form. Crippled with gout so that he could manage the stairs only
with help, he was well enough to bring his implacable righteousness to
bear with devastating effect. "Y^e P^r [president] ran himself into a great
heat & very indecent passion," he reported to Sharp, not without satisfaction. He demonstrated that all the instruments were his own private

property and thus not subject to the Visitors' authority. This nettled Newton, who said, "as good have no Observatory as no Instruments." Flamsteed complained of the publication of his catalogue, that they robbed him of the fruits of his labors – not an excessive description of the edition.

[A]t this he fired & cald me all the ill names Puppy &c. that he could think of. All I returnd was I put him in mind of his passion desired him to govern it & keep his temper. this made him rage worse, & he told me how much I had receaved from ye Goverm^t in 36 yeares I had served. I asked what he had done for ye 500^lb per Annum y^t he had receaved ever since he setled in London. this made him calmer but finding him goeing to burst out againe I onely told him: my Catalogue half finished was delivered into his hands on his own request sealed up. he could not deny it but said Dr Arbuthnot had procured ye Queens order for opening it. this I am persuaded was false, or it was got after it had been opened. I sayd nothing to him in return but with a little more spirit then I had hitherto shewd told them, *that God* (who was seldom spoke of with due Reverence in that Meeting) *had hitherto prospered all my labours & I doubted not would do so to an happy conclusion,* took my leave & left them.

"God forgive him," he added ineffably to his account of the same meeting in his autobiography. "I do."

Meanwhile Flamsteed was not idle. He had told Arbuthnot that he would print his catalogue as he wanted it at his own expense, and this he set out to do. In hot haste, he put the catalogue into final form, commenced printing in the summer of 1712, and had it done by the end of the year. Not yet satisfied, he began printing his observations with the mural arc. At this point, fate stepped in. Queen Anne died in 1714. The Tory government fell, and the Whigs returned to power. The death of Halifax in the summer of 1715 removed Newton's chief contact with the new regime, whereas Flamsteed knew the Lord Chamberlain, the Duke of Bolton. Someone near the Lord Chamberlain indicated to Flamsteed that if he wished he could get the copies of the *Historia coelestis* that had not been distributed. He did wish. On 30 November 1715, Bolton signed a warrant to Newton, the other referees, and Churchill to deliver the three hundred remaining copies of the work to the author. The referees replied that their authority over the work had ended with the queen's death. Although Flamsteed had received a gratuity of £125, they added disingenuously, and had caused a great deal of trouble by handing in imperfect copy, "yet we designed to have begged of her Majestie the Remainder of the Copies for him." No large degree of cynicism is required to doubt that statement. Finally, on 28 March 1716, Flamsteed got the copies into his hands. He separated out the catalogue and the hundred and twenty pages of excerpts from his observations with the mural arc and "made a *Sacrifice of them to Heavenly Truth.*" That is to say, he burned them. He devoted the brief time he had left to printing his observations with the mural arc, and he had virtually completed them when he died in 1719. His two former assistants, Joseph Crosthwait and Abraham Sharp, oversaw their completion and the printing of the Prolegomena and other material that went with the catalogue in volume 3 of the new work. In 1725, the *Historia coelestis britannica* finally appeared in three volumes: volume 1, the early observations salvaged from the 1712 edition; volume 2, the complete observations with the mural arc; volume 3, the catalogue and attendant materials. This was substantially the publication Flamsteed had always planned, and it is the form in which the *Historia* is known and honored today. Eventually, Flamsteed's widow and another assistant, James Hodgson, even published a brief version of the charts of the constellations, the *Atlas coelestis.* Newton's effort to subject Flamsteed to his will ended in complete failure, and the only solace he could find was the presentation of a copy of the 1712 edition, bound in red leather and gilt, to the Royal Society in 1717, a futile act of defiance which could not reverse his defeat. Flamsteed also found it impossible permanently to maintain his posture of lofty dignity. In June 1716 he told Princess Caroline, who was visiting his Observatory, that Newton was a great rascal who had stolen two stars from him. Unfortunately, the princess could not restrain herself from giggling.

The publication of the *Historia coelestis* was the most unpleasant episode in Newton's life. It was important both for the light it cast on his character and for what it revealed of his relations with the British scientific community. As for Newton, impatience with contradiction, which manifested itself in the young man in a readiness to throw caution to the winds in challenging established authorities such as Hooke, had become in his old age a tyrannical will to domineer, an unlovely trait which one cannot ignore. Perhaps the most interesting facet of the Flamsteed episode, however, was its revelation of the extent to which Newton had failed to make himself the dictatorial voice of British science. In 1709, two positions needed to be filled, the minor post in the Mathematical School at Christ's Hospital vacated by Samuel Newton's resignation, and the Savilian chair of astronomy at Oxford left empty by Gregory's death. Newton recommended William Jones for the first. Although we do not know that he meddled with the Oxford appointment, Halley apparently pushed John Keill, who had already established his credentials as a Newtonian. Neither one obtained the appointment in question. Flamsteed's former assistant, James Hodgson, became the mathematical master at Christ's Hospital. John Caswell was appointed Savilian professor of astronomy, to Flamsteed's delight. When Caswell died three years later, Keill did succeed him.

One must not imagine that Flamsteed challenged Newton's position in the British scientific world. No one challenged it. He dwelled on a level apart, and by his intellectual attainment alone he should have dominated it more than he did. The annual elections at the Royal Society revealed as much as the two positions he failed to fill. Year after year, others obtained more votes for membership on the council than Newton did. The minutes recorded fifty-odd members present in 1714; Newton received forty-five votes. In 1715, forty-nine members were present; Newton received thirty-five votes. William Derham returned to the council with forty-six. Newton did not attend the 1716 election. In 1723, when his senescence may have become a factor, only one of the eleven members of the council elected to continue received fewer votes. Year after year, approximately a fifth or a quarter of the members voted against him. The membership was made up mostly of nonscientists, of course, and elections had nothing to do with scientific achievement. We can take the elections as a rough measure of the extent to which Newton's despotic domination of the society alienated those who would not have hesitated to acknowledge his intellectual primacy.

12

The Priority Dispute

WELL BEFORE THE CONTESTED EDITION of Flamsteed's *Historia coelestis* in 1712 brought that episode to a temporary conclusion, two new concerns, which would dominate Newton's life for more than five years, had imposed themselves upon him. In 1709, work began in earnest on a second edition of the *Principia.* In the spring of 1711, a letter from Leibniz to Hans Sloane, secretary of the Royal Society, inaugurated a heated controversy over claims of priority in the invention of the calculus. Moreover, a fourth problem of great import for Newton was also taking form. Already an ugly scene with Craven Peyton, the warden of the Mint, had signaled a deterioration of their relations which culminated in a major crisis in the Mint in 1714, when the battle with Leibniz was reaching its highest pitch. The Mint was the bedrock on which Newton's existence in London stood. Trouble there had to affect his whole life. In its intensity, the period from 1711 to 1716, succeeding more than a decade of relative calm, matched the great periods of stress at Cambridge, when his relentless pursuit of truth stretched him to the limit. The coincidence of these events, the demands they placed on Newton, may help to explain the furious episode with Flamsteed at Crane Court on 26 October 1711 and much else from these years not yet mentioned.

The second edition of the *Principia* began first. Newton had been talking about it almost from the day the first edition appeared, first with Fatio, then with Gregory. With the move to London active plans languished for a time. They never died, however, and ultimately, on 25 March 1708, in some excitement, Gregory reported that the new edition was finally in the press in Cambridge.

There were certainly sound reasons not to put off the new edition any longer. Copies of the first edition were hard to come by and consequently expensive. What swayed Newton, however, were less these considerations than the manipulations of Richard Bentley, the academic entrepreneur now installed as master of Trinity College. Bentley had long made it his business to cultivate Newton and now aggressively maneuvered him into acquiescence in an edition which Bentley himself published through the university press.

Bentley called upon a young fellow of Trinity to oversee the edition. Roger Cotes, twenty-seven years old in 1709, was one of Bentley's followers in his struggle to inject new life into seniority-ridden Trinity. Through the tutoring of his uncle, the Reverend John Smith (father of Robert Smith, who would succeed Cotes at Cambridge and establish a modest reputation as a scientist), Cotes had made great progress in mathematics before his admission to Trinity in 1699. Bentley spotted him early and tapped him in 1705 for the newly created Plumian professorship of astronomy before Cotes had yet incepted M.A. Needless to say, Cotes gave his full support to Bentley's attempted reforms of college life. In 1709, Bentley, who could command rather than ask Cotes, put him in charge of the edition. It was October before Cotes received the manuscript for roughly the first half of the book and a reply to his letter of August, which had indicated that he had been checking some of the demonstrations.

I would not have you be at the trouble of examining all the Demonstrations in the Principia. Its impossible to print the book wthout some faults & if you print by the copy sent you, correcting only such faults as occurr in reading over the sheets to correct them as they are printed off, you will have labour more then it's fit to give you.

Cotes had his own ideas of what his task entailed, and before he was done, he forced on Newton detailed reconsideration of the work far beyond what Newton had had in mind.

All of this came later, however. Book I involved little serious alteration. Cotes found nothing on which to comment, and it moved forward swiftly. No letters between Newton and Cotes during the following six months survive, and from the content of Cotes's letter on 15 April 1710 it is unlikely that any were sent. At that point they had completed page 224 – well into Book II and nearly half of the edition's ultimate 484 pages – and they were approaching the end of the copy Newton had supplied. If Book

I had presented few problems, Book II, on motions in and of fluid media, offered more. Cotes discussed several points in the Scholium to Proposition X. The letter revealed his careful attention to the technical details of a difficult text, though as it happened – a matter of some significance – he failed to find a major error in the proposition itself. Subsequent letters, which now became quite frequent, bombarded Newton with problems in the first half of Book II to which he did not immediately respond. It was a subject he confessed, "from wch I have of late years disused myself . . . ." Though he accepted some of Cotes's corrections, he rejected others, only to have Cotes return again to the attack. This was not a treatment to which Newton had accustomed himself over the years; and even though he had finally to acknowledge Cotes's reasoning, he clearly did not enjoy being made to run the gauntlet.

Mr Professor

I have reconsidered the 15th Proposition with its Corollaries & they may stand as you have put them in your Letters.

By the middle of May, when a problem in Section IV caused a delay, the press caught up with the correction of the text and came briefly to a halt. After they set Section IV to rights, Section VI, the beginning of Newton's theory of the resistance of fluids to the motion of projectiles, stalled both them and the press again. "You need not give your self the trouble of examining all the calculations of the Scholium [at the end of the section]," Newton advised Cotes. "Such errors as do not depend upon wrong reasoning can be of no great consequence & may be corrected by the Reader." Cotes gave himself the trouble nevertheless. In a short time they solved the problems of Section VI, and on 30 June Cotes announced that the press had set all the copy Newton had furnished and that he was going home to Leicestershire for six weeks for a vacation. Section VI of Book II had been completed. In nine months, they had set 296 pages, not far short of two-thirds of the total.

The correspondence of the previous three months had been a new experience for Newton. Unwillingly at first, he had allowed Cotes to draw him into a genuine scientific exchange which finds no parallel, except for later discussion with Cotes, in the whole of his correspondence. His letters at first were brusque and cold, even to the point of being curt. By June, however, he was beginning to enjoy the discussion as Cotes exposed him again to the original excitement of his great work. He did not repeat the derisory greeting to "Mr Professor." Indeed, he thanked Cotes more than once for his corrections, and his reply to the letter of 30 June, besides its promise to have the rest of the corrected text ready, concluded with a warm salutation.

I am wth my humble service to your Master [Bentley] & many thanks to your self for your trouble in correcting this edition

S^r

Your most humble servant
Is. Newton.

In fact Cotes extended his vacation to more than two months, finally returning to Cambridge early in September. True to his promise, Newton immediately dispatched a further batch of the text that included the rest of Book II and Book III as far as Proposition XXIV. It began with Section VII, the heart of Newton's theory of the resistance of fluids – the topic, he had told Gregory, that was giving him the most trouble. No part of the *Principia* had been more imperfect, and now the press ground to a halt for nine months as Newton and Cotes wrestled with its revision, and even then it hardly moved for the rest of 1711.

❧

Meanwhile, the letter from Leibniz had arrived. As Newton contemplated its implications, it began to dominate his consciousness to the point of virtually sweeping every other topic out of his mind. Documents of every sort from the following years tend to be interrupted by furious paragraphs against Leibniz, as Newton, in his typical way, honed his prose with infinite care to razor sharpness. The very completion of the second edition seems in the circumstances something of a miracle, perhaps possible only to the extent that the battle spread out beyond the field of mathematics to cover natural philosophy as well, so that the *Principia* sustained one part of the front.

As we have seen, the storm that burst in 1711 had been gathering for many years, indeed from the moment in 1684 when Leibniz chose to publish his calculus without mentioning what he knew of Newton's progress along similar lines. The publication in 1699 of volume 3 of Wallis's *Opera,* with the full texts of the two *Epistolae* of 1676 plus other letters that testified to Newton's progress by the year 1673 and Leibniz's status in

1674 and 1675, all set in print with Newton's aid and cooperation, was the decisive event which made the public dispute inevitable. Before 1699, Newton had alluded to the correspondence of 1676 in the *Principia* and in the truncated "De quadratura curvarum" published in Wallis's volume 2 in 1693. Only Newton and those in communication with him – and Leibniz – understood the allusions. No one on the Continent knew the extent of Newton's communication in 1676. Pierre Varignon interpreted the fluxion scholium in the *Principia* as Newton's acknowledgment of Leibniz's invention of the calculus. There is every reason to think that his reading of it was typical. When Johann Bernoulli saw the piece in volume 2 of Wallis's *Opera*, he suggested that Newton had concocted it out of Leibniz's papers. To be sure, Bernoulli repeated that charge later, but it is equally true that when he saw *Commercium epistolicum* in 1712, with the same letters Wallis printed plus some more, he advised Leibniz that his best defense was to prove that the letters had been altered. It does not appear that Wallis's *Opera* circulated widely on the Continent. The *Acta* (probably Leibniz writing anonymously) reviewed the volume as though the letters devoted themselves mostly to the further celebration of Leibniz's early genius. Nevertheless, in 1699 the full text of a correspondence that Leibniz had concealed became public knowledge, and from that time on Leibniz's conduct changed.

In the same year 1699, Fatio de Duillier made his last significant impression on Newton's life with the publication of a mathematical tract, *Lineae brevissimi descensus investigatio geometrica duplex (A Twofold Geometrical Investigation of the Line of Briefest Descent)*. Six years had passed since the rupture with Newton. We have no reason to think that Newton had any hand in the treatise. Leibniz, who knew of their former intimacy, undoubtedly thought Newton did. In any event, Fatio had gone out of his way to insult Leibniz and had passed beyond insult to something much more. Fatio had invented his own calculus beginning in 1687, he stated, and owed nothing whatever to Leibniz, who would have to pride himself on his other disciples but not on Fatio.

However, driven by the factual evidence, I recognize that Newton was the first and by many years the most senior inventor of this calculus: whether Leibniz, the second inventor, borrowed anything from him, I prefer that the judgment be not mine, but theirs who have seen Newton's letters and his other manuscripts. Nor will the silence of the more modest Newton, or the active exertions of Leibniz in everywhere ascribing the invention of the calculus to himself, impose upon any person who examines these papers as I have done.

Leibniz both reviewed the publication anonymously and published a signed reply to the charge in the *Acta*. In denying it, he carried his silence in regard to the correspondence in 1676 a step closer to outright falsehood by asserting that in 1684, when he first published his method, he knew only that Newton had a method of tangents.

Nor was Leibniz solely the passive object of aggression from England. In 1699, he also published an attack on David Gregory's attempted demonstration of the catenary, in which he slyly implied that the fault of the demonstration followed from the deficiency of the fluxional method.

In this atmosphere of rising suspicion and hostility, Newton finally decided that the time had come to publish a fully fleshed specimen of his mathematics, the two treatises appended to the *Opticks*. His introduction made it clear that he did not publish them merely as a whim. "In a Letter written to Mr. Leibnitz in the year 1676, and published by Dr. Wallis [he stated in the preface], I mention'd a Method by which I had found some general Theorems about squaring Curvilinear Figures . . . ." Thus he contrived to imply that "De quadratura," actually composed in the 1690s, stemmed from the early 1670s, a claim on a par with Leibniz's reply to Fatio about what he knew in 1684. The introduction to "De quadratura" addressed itself no less to Leibniz and fortunately stuck closer to the truth. "I gradually fell upon the Method of Fluxions which I have used here in the Quadrature of Curves," he asserted, "in the Years 1665 & 1666." An anonymous review of "De quadratura" appeared in the *Acta eruditorum* for January 1705. Newton later claimed that he first saw the review in 1711, a claim there is some reason to question. Whenever he saw it first, he never doubted that Leibniz wrote it. As we know from Leibniz's papers, he was correct. In describing the content of the work, the review transposed it into the language of the differential calculus, invented by Mr. Leibniz.

Instead of the Leibnizian differences, therefore, Mr. Newton employs, and has always employed, *fluxions, which are almost the same as the increments of the fluents generated in the least equal portions of time*. He has made elegant use of these both in his *Mathematical Principles of Nature* and in other things published later, just as Honoré Fabri in his *Synopsis geometrica* substituted the progress of motions for the method of Cavalieri.

Although he never acknowledged his authorship, Leibniz did argue that the passage did not imply plagiarism by Newton. To be sure, it was artfully worded, but only a dull ear can miss the intended ring of the words. Among those with sharper ears was John Keill, who published a paper on centrifugal forces in issue no. 317 of the *Philosophical Transactions* for September and October 1708. Toward the end of his paper, Keill inserted a blunt rejoinder to Leibniz's insinuation.

All of these [propositions] follow from the now highly celebrated Arithmetic of Fluxions which Mr. Newton, beyond all doubt, First Invented, as anyone who reads his Letters published by Wallis can easily determine; the same Arithmetic under a different name and using a different notation was later published in the Acta eruditorum, however, by Mr. Leibniz.

Journals did not circulate with lightning speed in the early eighteenth century. Although the issue was published in 1709, presumably early, Leibniz did not see Keill's article for some time. When he finally did see it, he complained to the Royal Society, in the letter that arrived in March 1711.

What Keill said was no more than the common currency that passed through the whole learned world of Great Britain. Undoubtedly he merely repeated what he had heard at Oxford from Wallis, Gregory, and Halley, not to mention Fatio, who was there during the early years of the century. In the 1690s Gregory had seen the correspondence that would come out in Wallis's volume and had formulated for himself the argument the letters made for Leibniz's plagiarism. "These letters are to be printed," he concluded, "in the folio that Dr. Wallis is now aprinting, in the order of their dates, without any notes or commentaries or reflections: but let the letters themselves speake." Keill chanced to become the vehicle of Newton's revenge. It could easily have been someone else.

With Leibniz's letter complaining of Keill the covert preliminaries came to the end, and the open battle long in preparation began at last. My account of the controversy has nothing to do with the issue of priority. As far as I am concerned, that issue has been settled by examination of the papers left by the two principals. Newton invented his fluxional method in 1665 and 1666. About ten years later, as a result of his own independent studies, Leibniz invented his differential calculus. Newton contended – with his endless redrafting, contended repeatedly – that second inventors have no rights. A more patently absurd claim could not have been advanced. The first inventor clutched his discovery to his breast and communicated almost nothing. The second inventor published his calculus and thereby raised Western mathematics to a new level of endeavor. Newton realized as much in the end, and a good half of his fury was flung at Leibniz as a surrogate for his former self who had buried such a jewel in the earth.

My account of the quarrel, I repeat, does not concern itself with the question of priority. Varignon once remarked that the glory of the invention was sufficient for both men. My account of the quarrel is the story of their inability to share it amicably. If the glory of the invention was enough for both, so was the blame for the contest. This they did succeed in dividing in rough equality. In Newton's case, the cluster of inhibitions and neuroses that had prevented him from publishing his method in the first place likewise held him back from a forthright assertion of his claims. For nearly thirty years, from the time of Leibniz's first paper, he confined himself in print to opaque references to a correspondence with Leibniz in 1676. Beyond a small circle of Newton's intimates, no one except Leibniz understood what he was hinting at. Meanwhile, he did grumble privately to his intimates, and through them he poisoned the minds of a whole generation of English mathematicians. Newton liked to say that he hated controversy and tried to avoid it. In the matter of the calculus such does appear to have been the case. Though he protested in private, he drew back time and again from a direct challenge to the perceived injustice, as though he understood too well where his passion would carry him if it were once unchained. Well he might hold back, for once Leibniz's letter aroused him he was fury incarnate.

As for Leibniz, his eagerness for acclaim, his need to gather and guard the intellectual capital that assured his livelihood, led him in 1684 to the fatal error of trying to seize the undivided credit for his stupendous invention by neglecting to mention the correspondence of 1676. To Newton, this was his original sin, which not even divine grace could justify. Leibniz's paper had said that his method reached to the most sublime problems, which could not be resolved without it or another like it [*aut simili*].

What he meant by the words AUT SIMILI was impossible for the Germans to understand without an Interpreter. He ought to have done Mr. Newton justice in plain intelligible Language, and told the Germans whose was the *Methodus* SIM-

ILIS, and of what Extent and Antiquity it was, according to the Notices he had received from England; and to have acknowledged that his own Method was not so ancient. This would have prevented Disputes, and nothing less than this could fully deserve the Name of Candor and Justice.

Perhaps such a statement by Leibniz in 1684 would have prevented the dispute. Perhaps nothing would have. At any rate he had not mentioned what he knew of Newton's work, and by 1699, when Wallis's volume appeared, it was already too late. By then he had posed for fifteen years as the sole inventor of the calculus. Even to his friends he had not mentioned that exchange. Fontenelle's *éloge* seventeen years later reveals the extent of his initial success in passing as the sole inventor. What Leibniz feared was the conclusion the English drew. Freely to acknowledge the exchange in its full scope at that late date might lead to suspicions about his own independent discovery. When one recalls Bernoulli's response to *Commercium epistolicum* – that Leibniz's best defense was to prove that the letters were frauds – we can appreciate his dilemma.

Nothing is more revealing of Leibniz's problem than his silence about "De analysi." As he and he alone knew, the *Epistolae* of 1676 were not the half of the matter. In London he had also read "De analysi," the treatise so prominent in the dispute because independent evidence established its date. Leibniz never breathed a word about "De analysi," and Newton, who for polemic effect hinted vaguely at Leibniz's stop in London, did not know he had seen it and initially never mentioned what he would have loved to shout across Europe. Only when Leibniz, shortly before his death, inadvertently revealed the extent of Collins's liberality in the fall of 1676, did Newton begin to realize that he might have seen the tract as well.

Leibniz's strategy in the dispute flowed from the mortal danger to which the venial sin of 1684 exposed him. Hence the response to Fatio – that in 1676 he learned only of a method of tangents – to which he later added that he also learned about infinite series. Hence the review of Wallis, which stressed the evidence of Leibniz's own accomplishment by 1676, leading on to the argument that Newton was the pupil in the exchange. Hence the curiously restrained review, also anonymous, of "De analysi," which never said a word about its date but presented an argument, or perhaps only a suggestion of an argument for the reader to flesh out, that "De analysi" merely employed Archimedes' method of exhaustion and Fermat's method of infinitesimals, which differed from the new concepts found in the calculus invented by the "illustrious Leibniz." His friend Christian Wolf, who also did not know that Leibniz had read "De analysi" in 1676, urged that he needed to carry that argument further and prove that the tract did not contain the algorithm of the calculus – another unconscious illumination of Leibniz's dilemma. What would Wolf have said if he had read "De analysi" and known that Leibniz had also read it in 1676? Leibniz chose rather to mention "De analysi" as little as possible.

Until the final act of the ensuing drama, at least, Leibniz was all urbane sophistication. He never missed a chance to praise Newton in public even while he attacked him anonymously and by indirect insinuation. In 1701, Sir A. Fontaine found himself at dinner with Leibniz at the royal palace in Berlin. When the queen of Prussia asked him his opinion of Newton,

Leibnitz said that taking Mathematicks from the beginning of the world to the time of Sr I. What he had done was much the better half – & added that he had consulted all the learned in Europe upon some difficult point without having any satisfaction & that when he wrote to Sr I. he sent him answer by the first post to do so & so & then he would find it out.

Two years earlier he had implied, anonymously, that Gregory's mistake with the catenary flowed from the defects of Newton's method.

⁂

Newton also hid behind a shield of anonymity and spoke through the mouth of another. It has become fashionable of late to push the blame for the controversy onto Keill. No doubt he did nothing to calm the storm, but Keill's belligerence was also Newton's style. If Newton were going to speak through the mouth of another, he could not have found an instrument more fit. Leibniz always understood that Keill was Newton's mouthpiece. A man of the world who would have understood a response of overt praise and covert thrust, Leibniz gasped in astonishment when he found that the author of the *Principia* and *Opticks* was a wild bull whose only impulse was to lower his head and charge – and that he had locked himself into the same pen with this embodiment of rage.

The goad that finally incited Newton was Leibniz's letter to the Royal Society dated 4 March 1711 in complaint of Keill's "impertinent accusation." Protesting his innocence, Leibniz applied to the Royal Society for a remedy, to wit, that Keill publicly testify that he did not mean to charge Leibniz with what his words seemed to imply. Newton presided at the meeting of the Royal Society on 22 March when the letter was read. Two weeks later, the Royal Society took up the matter again, and Newton, from the chair, presented his version of the invention of the calculus. When the minutes were read the following week, Newton amplified his remarks, mentioning "his Letters many years ago to M^r. Collins about his Method of Treating Curves &c . . . ." The society asked Keill, who was present again, to draw up a paper that asserted Newton's rights.

A month and a half went into the preparation of Keill's reply, the period during which Cotes waited for Newton's final resolution of Proposition XXXVI, on fluid resistance. We do not have manuscript evidence of Newton's extensive participation in the letter's composition, but everything about it – its intimate knowledge of Newton's early papers and correspondence, details of its argument about Leibniz's progress as revealed by his correspondence in 1675 and 1676, above all its style, which treated the issue as a historical question to be settled by empirical evidence from the manuscript record – cries aloud of the hand that shaped it. The Royal Society, with Newton presiding, heard the finished letter on 24 May and ordered that it be sent to Leibniz but not published in the *Philosophical Transactions* until Leibniz's answer showed that he had received it. Newton even drafted Sloane's covering letter. In fact, Keill's reply to Leibniz's complaint never appeared in the *Philosophical Transactions;* when Leibniz's response arrived, it gave rise to a more extended publication.

Leibniz devoted some thought to his answer, which he did not send until 29 December and which the Royal Society did not receive until 31 January 1712. Though he could not have failed to understand that Newton, the president of the Royal Society, stood behind a letter transmitted to him by the society, Leibniz chose carefully to distinguish Newton from Keill, whom he treated as an upstart before whom he had no need to justify himself. As a member of the Royal Society, he appealed to it for justice. To this appeal Newton lit upon an unexpected riposte. Leibniz had thrown himself on the justice of the society. Very well, let the society sit in judgment on the question. On 6 March 1712, the Royal Society appointed a committee to inspect the letters and papers relating to the matter: Arbuthnot, Hill, Halley, Jones, Machin, and Burnet, to which they later added Robartes, DeMoivre, Aston, Taylor, and Frederick Bonet, the minister in London of the king of Prussia. Newton liked to refer to this committee as "numerous and skilful and composed of Gentlemen of several Nations . . . ." Rather it was a covey of his own partisans to which Herr Bonet, to his undying shame, allowed himself to be co-opted to provide some minimal veneer of impartiality.

Neither Herr Bonet nor the others found the task difficult. Newton had already done the spadework in a year of intensive investigation. Only a few years earlier, William Jones had come into possession of the papers of John Collins. The full advantage of this stroke of luck now emerged. Together with Oldenburg's correspondence in the records of the society, Collins's papers provided the factual foundation of a report which did not need to call upon Newton's own papers at all. Leibniz later complained – with full justice – that the committee conducted a judicial proceeding without informing him of the fact or allowing him to present evidence. The committee did not call upon Newton to present evidence either. It did not need to. Newton carried out its investigation, arranged its evidence, and wrote its report, which explains why the committee was able to submit the report, which presumed to survey the whole history of the calculus, on 24 April, a full month and a half after its appointment. The final three members had been appointed on 17 April, one week earlier. The procedure may explain why the committee did not sign the report, but Newton's creatures were too tame to balk at submitting it. Of his own central participation there can be no doubt. Beyond the extensive manuscript remains of his researches, which went into the volume later published, his draft of the report still exists. In his later "Account of *Commercium epistolicum*," Newton waxed indignant over Leibniz's claim to have invented his calculus before he received the letters of 1676. "But no Man is a Witness in his own Cause," he thundered. With his own words he passed judgment on himself.

Not surprisingly, the committee, or court, found in Newton's favor, in a condemnation of Leibniz beside which Keill's paragraph looks like praise. By the letters and papers they found

That the Differential Method is One and the same with the Method of Fluxions Excepting the name and Mode of Notation . . . and therefore wee take the Proper Question to be not who Invented this or that Method but who was the first Inventor of the Method, and wee beleive that those who have reputed M^r. Leibnitz the first Inventor knew little or Nothing of his Correspondence with M^r. Collins, and M^r. Oldenburg Long before, nor of M^r. Newtons haveing that Method above Fifteen Years before M^r. Leibnitz began to Publish it, in the Acta Eruditorum of Leipsick.

For which Reasons we Reckon M^r. Newton the first Inventor and are of Opinion that M^r. Keill in Asserting the same has been noways Injurious to M^r. Leibnitz and wee Submitt to the Judgment of the Society whether the Extract of the Letters and Papers now Presented Together with what is Extant to the same Purpose in Doctor Wallis's third Volume may not deserve to be made Publick.

The resulting volume, *Commercium epistolicum D. Johannis Collins, et aliorum de analysi promota (The Correspondence of the Learned John Collins and Others Relating to the Progress of Analysis)*, appeared early the following year. Not much needs to be said about *Commercium epistolicum*. The overwhelming bulk of the volume consisted of letters and papers and extracts chronologically arranged, leading up to the judgment of the committee, which it printed in full in its original English. Footnotes of the most partisan kind, pointing out the fluxional method in Newton's papers and letters and denigrating the letters of Leibniz, provided a running commentary. The passion Newton invested in the argument as he took his revenge burst out in mathematical notes where one scarcely expects to find it. Leibniz's letter of 12 July 1677 remarked that he had found some of Newton's series among old papers of his own that he had forgotten. Newton's indignation erupted. Leibniz had received the series two years earlier, asked for the method behind them, received it from Oldenburg, had trouble understanding it, and now found that he had discovered it earlier himself. So it had been with other series. "Thus the method which earlier he wanted, asked for, received, and understood with difficulty, he discovered forsooth either first or at least by his own effort."

Commercium epistolicum was a brilliant exercise in partisan polemics which testified to Newton's continuing mental vigor as he approached the age of seventy. The total impact of the notes, the total impact of the whole volume in the absence of anything in Leibniz's defense, is devastating. Perhaps it is too devastating. Swept along by his own fury, Newton failed to recognize the utility of moderation. No doubt the volume informed a particular public of events that Leibniz had not been forward to advertise. It is not clear that it was convincing, however. Leibniz had made too deep an impression on the learned world to be hustled off the stage as a fraud at this point in his career. "Since it does not appear that M. Leibniz is satisfied with this decision," the *Journal des sçavans* commented dryly, "the public will no doubt receive further information on the subject from him."

&

The second edition of the *Principia* hardly moved during 1711 as the dispute with Leibniz, not to mention Flamsteed's *Historia coelestis* and a busy year at the Mint, distracted Newton. In June, the problems with Proposition XXXVI and Section VII, Book II, as a whole, which had stalled the press for nearly a year, were finally resolved. Bentley brought all the rest of the copy from London, and the press seemed to move once more. Unfortunately, Cotes found new problems with Proposition XLVII in Section VIII, on the propagation of pulses such as sound through elastic media; and another eight months elapsed before they were settled. As things stood in February 1712, the edition, which had progressed swiftly through 296 pages in the first nine months, had managed to add a scant 40 more in the following nineteen. At this point, however, Newton set his mind seriously on his masterwork for the last time, and the edition proceeded to its conclusion, though not without two more minor interruptions.

The priority dispute with Leibniz, which had spilled over into the philosophic differences that separated the two men, influenced aspects of the second edition. Newton's reply to the philosophic criticisms from the Continent and from Leibniz, which objected vehemently to the whole concept of forces that act at a distance, contrasted his experimental philosophy with the hypothetical philosophy of his antagonists. He did not, he insisted, attempt to teach the causes of phenomena except insofar as experiments revealed them. He did not wish to fill philosophy with opinions that experiments could not prove. His critics found it a fault that he did not offer some hypothesis about the cause of gravity, as though it were an error not to dilute demonstrations with speculations. Thus the major thrust of the new edition was to further emphasize that feature of the *Principia*, which had opened itself to him in his final expansion of the work. The second edition changed little in Book I, where the consideration of the gross features of the universe led to the recognition of inverse-square attractions. The classic demonstrations of the dynamic foundation of Kepler's laws required no revision. He concentrated rather on further enhancing the work's derivation of the quantitative details of physical phenomena, that revolutionary feature which endless pejorative incantations against occult qualities could not conjure away. Such was the issue behind Section VII of Book II, on which the edition lay marooned for a year. Such were the major revisions of Book III, the final polishing of which still lay ahead.

Book III opened with a declaration of philosophic principle in a new rule of reasoning, Rule III, perhaps his most important statement of epistemology.

The qualities of bodies, which admit neither intensification nor remission of degrees, and which are found to belong to all bodies within the reach of our experiments, are to be esteemed the universal qualities of all bodies whatsoever.

The extended discussion of Rule III, addressed to Cartesians, to mechanists in general, and to Leibniz in particular, contrasted his empirical experimental philosophy with hypothetical philosophy. "For since the qualities of bodies are only known to us by experiments, we are to hold for universal all such as universally agree with experiments . . . We are certainly not to relinquish the evidence of experiments for the sake of dreams and vain fictions of our own devising . . . ." Because experiments and astronomical observations show that all bodies about the earth gravitate toward it in proportion to their quantity of matter, that our sea gravitates toward the moon, and all planets toward one another, "we must, in consequence of this rule, universally allow that all bodies whatsoever are endowed with a principle of mutual gravitation."

Cotes had no comment to make on Rule III. The intense correspondence about the details of Book III that now commenced between the two focused on emendations intended to emphasize the success of Newton's science in explaining the phenomena of nature with quantitative precision. One such case had already arisen in Book II in connection with Newton's derivation of the velocity of sound from first principles of dynamics. In Book III the correlation of the moon's orbit with the measured acceleration of gravity on the surface of the earth, and the derivation of the precession of the equinoxes, were very similar. In all three cases, edition one had been content to rest the case on correlations that were only approximate. Edition two, for polemical purposes, contrived to manipulate essentially the same body of empirical data to give the illusion of accuracy to better than one part in a thousand.

He did not succeed equally with lunar theory, which occupied Propositions XXV–XXXV of Book III. It was probably inevitable that nothing
important would come of Newton and Cotes's exchange on this subject.
Newton had exhausted himself on lunar theory in the 1690s and was
incapable now of seriously amending it. The changes in lunar theory in
the second edition, especially the new scholium that concluded it, had
been worked out at the earlier time, and they did not approach the high
expectations Newton had once entertained.

Their correspondence took a sudden turn in April, when Cotes suggested that the two of them undertake a volume of Newton's mathematics
after the *Principia* was done. He had found some errors in the treatises
already published that they could correct. Cotes had presumed too far.
For two-and-a-half years the two men had engaged in intellectual exchange, an extremely intense one during the last three months. Cotes's
letter of 26 April terminated it. Newton never even answered to the
proposal Cotes made, and he let three letters accumulate before he replied at all. Though Newton did write twice during May, both letters were
brief, even curt. He did not write again for three months while Cotes
waited impatiently for an answer to questions on lunar theory that had
been left dangling. Among Newton's papers there is a draft of a preface
for the edition, with a handsome tribute to "the very learned Mr Roger
Cotes," his collaborator, who corrected errors and advised him to recon-
sider many points. At some point, probably about this time, he suppressed
it, and the new edition appeared without any mention of Cotes other than
his name at the end of the preface he wrote.

Three weeks after they finished with the moon, on 14 October another
letter from Newton abruptly announced, "There is an error in the tenth
Proposition of the second Book, Prob. III, wch will require the reprinting
of about a sheet & an half." The origin of this blunt message was a visit of
Nikolaus Bernoulli, the nephew of Johann Bernoulli, to London that fall.
Nikolaus informed Newton that his uncle had discovered an error in
Proposition X, on the motion of a projectile under uniform gravity
through a medium that resists in proportion to velocity squared. Bernoulli
had discovered the mistake by 1710. For him it became evidence that
Newton had not understood second derivatives when he wrote the *Principia,* a matter of obvious relevance to the priority dispute. It seems clear
from subsequent correspondence that Bernoulli, knowing a second edi-

tion was in progress, intended to delay revelation of the error, in the expectation that the same defect, uncorrected, would not only display his own brilliance but would also become an effective argument that the Leibnizian calculus was a different and more powerful instrument than Newton's fluxional method. He had not reckoned on his nephew's visit to London.

Although Newton did not then know of the trap Johann Bernoulli was trying to set, he saw immediately that much more than a particular result was at stake. He had just told Cotes that the public would have to take the lunar theory as it was. He had no intention that the public receive, much less take, Proposition X, an apparent demonstration of mathematical weakness, as it was. In October 1712, he was less than three months removed from his seventieth birthday, not anyone's prime age for mathematics. Nevertheless, he not only fell to on the problem with a will, quickly consuming some twenty sheets of paper in its consideration, but he found the source of his error – which Bernoulli, who had found the error by solving the problem by a different method, had not been able to do. He dashed off a letter to Nikolaus Bernoulli, sending him the corrected proposition and asking him to send it to his uncle. On 14 October, Newton told Cotes that there was an error and that the proposition would need to be corrected. By 18 October, according to DeMoivre, he had amended it. Nevertheless, he let Cotes wait another three months before he sent the revised proposition. It was necessary to reprint one whole sheet, signature Hh, plus the final two pages of the previous sheet, signature Gg, which were inserted as a cancel pasted to the stub of the original page, telltale evidence of a late correction that Bernoulli did not fail to notice.

ॐ

To his brief letter that accompanied the revised Proposition X, Newton added another surprise for Cotes: "I shall send you in a few days a Scholium of about a quarter of a sheet to be added to the end of the book: & some are perswading me to add an Appendix concerning the attraction of the small particles of bodies." About the same time, Bentley told Cotes orally that he should compose a preface to the volume. Thus were conceived the two most visible additions to the second edition. Newton and Newtonians were highly aware of the mounting tide of criticism of his natural philosophy and its concepts of attractions and repulsions, a tide rising with the priority dispute. Both Cotes's preface and Newton's General Scholium replied to the criticism. Placed at the front and rear, like symbolic covers, they gave to the new edition a polemic tone in harmony with the modifications designed to enhance the work's appeal to natural philosophers.

The appendix on the attractions of particles fell by the way, but Newton sent the General Scholium in March. Despite the reference to an electric spirit in the final paragraph, the General Scholium opened with a ringing challenge to mechanical explanations of the heavens. "The hypothesis of vortices is pressed with many difficulties," he began. As he proceeded to detail them – comets, resistance to motion, and the contradiction, for vortices, between Kepler's second law and his third – the verb *pressed* hardly seemed adequate to express the desperate straits of vortical theory. In more general terms, the order of the cosmos is incompatible with mere mechanical necessity.

This most beautiful system of the sun, planets, and comets, could only proceed from the counsel and dominion of an intelligent and powerful Being. . . . This Being governs all things, not as the soul of the world, but as Lord over all; and on account of his dominion he is wont to be called *Lord God*, παντοκρατωρ or *Universal Ruler;* for *God* is a relative word, and has a respect to servants; and *Deity* is the dominion of God not over his own body, as those imagine who fancy God to be the soul of the world, but over servants.

Newton continued, expounding his conception of God and of absolute space and time as the consequences of His infinite extension and duration.

He is omnipresent not *virtually* only, but also *substantially;* for virtue cannot subsist without substance. In him are all things contained and moved; yet neither affects the other: God suffers nothing from the motion of bodies; bodies find no resistance from the omnipresence of God.

God is devoid of all body and ought not to be worshiped through any material image. We have ideas of His attributes; we cannot know His substance. We know Him by His works; we admire Him for His perfections; "but we reverence and adore Him on account of His dominion: for we adore Him as his servants; and a god without dominion, providence, and final causes, is nothing else but Fate and Nature."

So far, he concluded, he had explained the phenomena of the heavens by the force of gravity but had not shown the cause of the force – a dubious statement after what he had just said about the dominion of God. He explained what the cause must account for: the action of gravity, not in proportion to the surfaces of bodies "(as mechanical causes used to do)" but in proportion to quantity of matter, its penetration to the very center of all bodies without diminution, its propagation to immense distances decreasing in exact proportion to the square of the distance. "But hitherto," he proceeded, in one of his most frequently quoted passages, "I have not been able to discover the cause of those properties of gravity from phenomena, and I feign no hypotheses . . . And to us it is enough that gravity does really exist, and act according to the laws which we have explained, and abundantly serves to account for all the motions of the celestial bodies, and of our sea." Composed virtually at the end of his active life, the General Scholium contained a vigorous reassertion of those principles which Newton had adopted in his rebellion against the perceived dangers of Cartesian mechanical philosophy. The same principles had continued to govern his scientific career as he followed the consequences of his rebellion into a new natural philosophy and a new conception of science.

As spring passed into summer in 1713, the edition finally approached completion. It was only on 30 June that Bentley announced it to Newton. "At last Your book is happily brought forth; and I thank you anew yᵗ you did me the honour to be its conveyor to yᵉ world."

Cotes received neither any payment for his labor from Bentley's profits from the edition nor even any word of thanks from Newton. As I have mentioned, Newton suppressed the preface that contained a tribute to him and expunged a reference to him from the text. Six months after the book had been published, he suddenly sent Cotes a list of Errata, Corrigenda, and Addenda which he apparently expected to have printed and bound with the volume. Somewhat chagrined, Cotes got his back up far enough to say that he had found several of the errata himself, "but I confess to You I was asham'd to put them in the Table, lest I should appear to be too diligent in trifles." He knew of at least as many more, he added, that Newton had missed. He did not get his back up far enough to substitute plain "you" for the divine mode of address, however.

ଈ

Meanwhile the dispute with Leibniz continued. On 29 July 1713, a sheet without the name of its author, its printer or even the city in which it was printed appeared and spread quickly through interested circles on the Continent. Known as the Charta volans or (as Newton anglicized it) flying sheet, it was Leibniz's reply to *Commercium epistolicum.* The Charta volans said little in detail about the correspondence of the 1670s. Rather it pointed out that Newton had published nothing on the calculus before Leibniz did and went on to assert that when the English began to ascribe everything to Newton, Leibniz, who had hitherto been inclined to believe Newton's claim of independent discovery, reexamined the matter more carefully and became convinced that Newton had developed the fluxional method in imitation of his calculus. In support of the last, the Charta quoted a letter of 7 June 1713 from a "leading mathematician" who expressed the opinion that in the 1670s Newton had invented only his method of infinite series. Apparently no one was ever deceived as to the author of the Charta volans. Bernoulli told Leibniz that it was openly attributed to him, and Newton gave two good reasons why Leibniz was the author to John Arnold, an English friend of Leibniz who relayed them to Hanover.

The "leading mathematician" quoted by the Charta volans was Johann Bernoulli. Though suspected, Bernoulli's authorship of the passage proved to be a better-kept secret. Written as soon as Bernoulli had seen a copy of *Commercium epistolicum,* the letter probably gave Leibniz his first account of the volume. It was a scathing account, as Bernoulli was outraged by the injustice of the whole proceeding. Nevertheless, he was not anxious to make his outrage public. "I do indeed beg you [he concluded the letter] to use what I now write properly and not to involve me with Newton and his people, for I am reluctant to be involved in these disputes or to appear ungrateful to Newton who has heaped many testimonies of his goodwill upon me." Leibniz indicated to Bernoulli that though he did not wish to embroil him in quarrels, "I expect from your honesty and sense of justice that you will as soon as possible make it evident to our friends that in your opinion Newton's calculus was posterior to ours, and say this publicly when opportunity serves . . . ." By immediately printing his letter in the Charta volans, albeit anonymously, Leibniz gave himself the means of making Bernoulli's support public should Bernoulli hold back, and in the end, as the dispute continued endlessly, he did make Bernoulli's authorship known.

With two men angry beyond willingness to hear reason, the dispute could only grow. The inaugural issue, for May and June 1713, of the *Journal literaire*, a new journal launched by a group of Dutch savants, carried an anonymous letter from London, written by Keill, which presented the Newtonian version of the priority question together with French translations of the report of the committee of the Royal Society and of Newton's tangent letter of December 1672. Leibniz felt he must answer. Hence the issue for November and December of the same journal printed a French translation of the Charta together with anonymous "Remarks" on the difference between Newton and Leibniz. In the "Remarks" Leibniz, who composed them, repeated at greater length the case he had made in his letters to Bernoulli, laying emphasis on the absence of the calculus from the *Principia* where it was needed and the lack of understanding shown when the work did try to use it. Newton's error on Proposition X and the proof it seemed to offer that he did not understand second-order differentials in 1687 loomed ever larger in an argument seriously short of empirical content. In due time Keill, who was known on the Continent with good cause as Newton's ape, responded to the "Remarks" with a paper that was largely Newton's composition.

As the dispute deepened, it also broadened to include more men. On the Continent, Bernoulli and Wolf labored on Leibniz's behalf, one more covertly than the other. In England, Newton co-opted not only Keill but also the Huguenot refugee DeMoivre, who became the translator of whatever went to the *Journal literaire* for publication. Bernoulli and DeMoivre struggled to maintain their personal correspondence, writing deliberately misleading letters which attempted to conceal their participation. Almost no British mathematician or natural philosopher remained unmoved by the cause.

Toward the end of 1714, worried that his version of the struggle had not reached its intended audience, Newton began to compose his own essay on the dispute, "An Account of the Book entituled *Commercium Epistolicum*." He could not, of course, put his name to it and appear openly in the lists. He could publish it anonymously in the *Philosophical Transactions* for January–February 1715, however, filling all but three pages of the whole issue. He could have DeMoivre translate it into French for publication in the *Journal literaire*. He could have a review of it sent to the *Nouvelles littéraires*. He could have the French version of the "Account" printed as a separate pamphlet. He could have copies of the pamphlet spread about the Continent by whatever means presented themselves. And if he could, he did. Later he also published it in Latin.

Leibniz, who soon learned that it was disastrous to challenge Newton on empirical questions about the historical record, always sought to push the discussion onto broader philosophical grounds. In November 1715, replying to a question about the theology of Samuel Clarke put to him by the princess of Wales, Leibniz sent a famous challenge, which could not be ignored.

Natural religion itself, seems to decay (in England) very much. Many will have human souls to be material: others make God himself a corporeal being. . . . Sir Isaac Newton says, that space is an organ, which God makes use of to perceive things by. . . . Sir Isaac Newton, and his followers, have also a very odd opinion concerning the work of God. According to their doctrine, God Almighty wants to wind up his watch from time to time: otherwise it would cease to move. He had not, it seems, sufficient foresight to make it a perpetual motion.

Leibniz addressed the challenge more to Clarke, who at that time was attending upon the princess assiduously, than to Newton, and Clarke (who was a dedicated Newtonian in any case) undertook to reply. The exchange lasted through five rounds, ten letters in all, each one longer than the last as every point made required more words from the other side for its refutation, until Leibniz's death finally terminated the correspondence. Leibniz's original letter defined perhaps the central issue, the divine governance of the universe. Inevitably the discussion also spread out into natural philosophy and embraced questions such as attractions and voids. Although it did not figure as a central issue, Leibniz could not bring himself to ignore Newton's conception of space as the sensorium of God. Once he saw it in Query 28 of the *Opticks*, it appeared to him an absurdity that cast doubt on Newton's competence in philosophy as a whole. "And so this man has little success with Metaphysics," he told Bernoulli in commenting on the passage. It remains an open question how extensively Newton participated in the composition of Clarke's side of the correspondence. Unlike the case of Keill's letters, extensive manuscript evidence of his active role does not exist. On the other hand, Clarke was one of his close adherents. He was the rector of the chapel in Golden Square of which Newton was a trustee, two Arians masquerading as orthodox.

Though he tried not to be diverted from the question of priority by philosophical issues, Newton did take up the differences between himself and Leibniz. The General Scholium to the *Principia* had already done so, and the "Account of *Commercium Epistolicum*" also closed with three pages devoted to a forceful exposition of the differences between their philosophies. Newton's philosophy was experimental, Leibniz's hypothetical. He quoted anew the various reservations he had inserted in the *Principia* and the *Opticks* about forces and their possible causes. "And after all this, one would wonder that Mr. Newton should be reflected upon for not explaining the Causes of Gravity and other Attractions by Hypotheses; as if it were a Crime to content himself with Certainties and let Uncertainties alone." Can the constant and universal laws of nature, whether derived from the power of God or from a cause yet unknown, be called miracles and occult qualities, that is, wonders and absurdities? Must all the Arguments for a God taken from the Phaenomena of Nature be exploded by *new hard Names*? And must Experimental Philosophy be exploded as *miraculous* and *absurd*, because it asserts nothing more than can be proved by Experiments, and we cannot yet prove by Experiments that all the Phaenomena in Nature can be solved by meer Mechanical Causes? Certainly these things deserve to be better considered.

The manifesto, written in the twilight of Newton's scientific career, voiced his continued support of the principles that had guided him to his revolution in natural philosophy.

"Mr. Leibniz is dead; and the dispute is finished." Newton received this word in December 1716 from Antonio Schinelli Conti, a Venetian who had become Newton's close friend in London during the previous year, who wrote from Hanover, where he had gone in the hope of meeting Leibniz. He had died on 4 November, well before Conti arrived. As to the second clause in his announcement, Conti could not have been more mistaken. The passions generated had reached a pitch that required another six years for their dissipation. Nevertheless, Leibniz's death removed the object of Newton's wrath, and with time even he tired of the repetition of stale taunts. If the dispute was not finished, at least its conclusion had been announced. Newton had probably celebrated his seventy-fourth birthday before Conti's message arrived. With it ended the last passionate episode of his life. Though he had more than ten years to live, they were, inevitably for a man of his age, years of decline.

13

Years of Decline

THE PRIORITY DISPUTE dragged on with diminished intensity for another six years and during that time continued to occupy a major part of Newton's consciousness. He had never been able to lay a project down easily. Wound up as tightly as he was now, and with his honor at stake, he could not put the dispute aside simply because his antagonist had died. It was 1723 when its final faint echo was Newton's refusal to reply to a letter from Bernoulli.

Among its other effects, the controversy served to remind Newton that he needed to give attention to his intellectual legacy. As a result, he devoted considerable attention during his old age to new editions of his works. In 1717, it was a new edition of the *Opticks*. He did not touch the body of the treatise, which continued to set forth conclusions as he had established them forty-five years earlier, but he composed a set of eight new Queries, which he inserted as Queries 17–24 between the original set of sixteen Queries in the first edition and the set of seven added to the Latin edition. Continuing a retreat from the radical stance of earlier years, he now postulated a cosmic aether to explain gravity. To be sure, the aether had so little in common with conventional mechanical fluids that the retreat was more apparent than real, and Newton may have intended it more as a sop than a concession. The aether, he said, was "exceedingly more rare and subtle than the Air, and exceedingly more elastick and active." He concluded indeed that the ratio of its elastic force to its density must be more than 490 billion times that of air. Was such a medium possible? It was, Newton argued, if one supposed that the aether, like the air, is composed of "Particles which endeavour to recede from one another . . . ." That is, Newton's new aether embodied the very

297

problem it seemed to explain, action at a distance in the form of a mutual repulsion between aethereal particles. Standing rarer in the pores of bodies than in free space, the aether caused the phenomena of gravity by its pressure.

The *Opticks* was in English, of course. Only a Latin edition could effectively reach the Continental audience, and in 1719 Newton published a second Latin edition that included the new Queries. In 1720 Pierre Coste's French translation was published in Amsterdam, just at the time when the first successful reproduction of Newton's experiments in France had raised interest in his theory of colors; and two years later a second French edition appeared in Paris. Meanwhile, in 1721, Newton issued a third English edition, which did not differ significantly from the second.

Newton also gave one last effort to the *Principia*. It is clear that he regarded the *Principia* rather than the *Opticks* as his masterwork. He had thrown the *Opticks* together largely from previously written papers in the 1690s and had scarcely touched the text, as opposed to the Queries, through three English and two Latin editions. In contrast, he worked over the *Principia* without end to hone its language to a perfect expression of his ideas. Perhaps the appearance of a reprint of the second edition in Amsterdam in 1723 stimulated Newton to put his plan for a new edition into action. Perhaps a serious illness in 1722 reminded him that he could not delay forever. We know only that printing of an edition more sumptuous than either of the others began in the fall of 1723. As editor, Newton had the services of a young member of the Royal Society, Henry Pemberton, who had recently returned to England after completing a medical degree in Leyden in 1719. In the fall of 1723, Pemberton addressed to him the first of thirty-one communications which stretched over the following two-and-a-half years while the edition passed through the press.

By universal agreement, Pemberton performed his function with less understanding and skill than Cotes had shown with the second edition, and he left little of himself impressed upon the work. His letters tended to be short and uninteresting, raising only small points of style. To be fair to Pemberton, it is necessary to recall that Newton was now more than eighty years old and no longer capable of the serious exchange that occurred with Cotes. On occasion Pemberton did try to press beyond details, and he brought up some basic issues that had been among the more hotly debated topics in the second edition: the flow of water from a tank, and aspects of the treatment of tides that bore on the ratio of the sun's force to the moon's and thus on the precession of the equinoxes. Newton refused to become engaged and simply swept the arguments aside. There is other evidence to confirm that Newton could no longer sustain a penetrating reexamination of his work.

The most important addition appeared near the beginning of Book III, a new Rule IV for reasoning in philosophy which carried on the argument with Leibniz. Among Newton's papers are a number of drafts that testify to plans which were at times much more extensive. Eventually he settled for a restrained statement from which he succeeded in eliminating the passion that had mounted anew within the octogenarian as he thought once more about Leibniz.

Rule IV. In experimental philosophy we are to look upon propositions inferred by general induction from phenomena as accurately or very nearly true, notwithstanding any contrary hypotheses that may be imagined, till such time as other phenomena occur, by which they may either be made more accurate, or liable to exceptions.

This rule we must follow, that the argument of induction may not be evaded by hypotheses.

Through 1724 and 1725 the edition made its slow but steady progress toward completion with none of the delays that stopped the press during the second edition. Newton dated the preface 12 January 1726. It was the last day of March when Martin Folkes presented a copy "richly Bound in morocco Leather" to the Royal Society in Newton's name. In all, 1,250 copies were printed, fifty of them on superfine paper. Newton intended at least some of these as presentation volumes. He gave one of them to his friend and associate of thirty years at the Mint, John Francis Fauquier. He was lavish in his concern for the Paris Académie; it received no fewer than six copies.

᠌᠌᠌ ૨▲

The mathematical and scientific works did not, however, exhaust the Newtonian legacy, and in his old age Newton turned again to the theological interests that had burned with consuming intensity during the years of his early manhood. Although he had never completely abandoned theological studies, a hiatus of about twenty years beginning with the publication of the *Principia* had interrupted their vigorous pursuit. Only a small number of theological papers can be placed with assurance in this period. In the years 1705–10, he returned to theology with renewed vigor, and for all his editions of the *Principia* and *Opticks*, theology was the primary occupation of his old age. Some of the papers, in which information on early heresies appears on the same page with material on heathen religions or the nature of Christ or the chronology of the prophecies, suggest that during his final years his mind was frequently a chaos of undifferentiated religious preoccupation. Newton's conclusions in theology had been as radical as his conclusions in natural philosophy. So far they had seen the light of day only within a limited circle of trusted confidants. As the years mounted and he became conscious that the inevitable end approached, Newton gave thought to his responsibilities. He said when he died "he should have the comfort of leaving Philosophy less mischievous" than he found it, Conduitt reported; "he might say the same of the revealed religion then mention his Irenicum his creed." What Conduitt did not mention were the factors that made it even more difficult for an old man fond of his position and respect to publish heretical views than it had been for a rebellious young don fifty years earlier.

Newton had been at work revising his interpretation of the prophecies and other theological works for about ten years when Caroline, princess of Wales, heard about his new principles of chronology in 1716. Interested, she summoned Newton and asked for a copy of what he had written. Newton never lightly surrendered one of his compositions. He had even less desire to hand over to the princess of Wales a treatise which might still have contained assertions heretical enough to secure his instant dismissal from the Mint. Well schooled as he was in the art of delay, he pleaded that the work was "imperfect and confus'd," but he knew very well that one did not dally with a royal command. In haste he drew up an "Abstract" of his chronology, what was later called the "Short Chronology," which put the work into "that shape the properest for her Perusal . . . ," and delivered it to the princess in a few days. By themselves, cut off from the "Origines" which was their source, there was nothing very novel in the ideas the "Abstract" presented and nothing to excite odium. By disguising radical theology as chronology, Newton had made it safe enough even for royal consumption.

Newton had not heard the last of the "Abstract," however. A copy made its way to France where a translation, entitled *Abrégé de la chronologie,* was published in 1724, together with a refutation composed by Nicholas Fréret of the Académie des Inscriptions et Belles-lettres. Nor were the *Abrégé* and the "Remarks" the end of the business. Father Etienne Souciet, another expert on ancient chronology, published five dissertations against Newton's system of dates. Initially Newton, who was now an old man, undertook to ignore the attacks in order to avoid another controversy. Eventually he changed his mind and decided that he must defend himself by bringing his full treatise on chronology out. He was at work revising it when he died. We should not from that conclude that the work was unfinished. Endless revision, however minute the alterations, was Newton's fate. Conduitt published the full volume, *The Chronology of Ancient Kingdoms Amended,* in 1728, the year after Newton's death.

At one point Newton had envisaged a bolder project than the carefully sanitized *Chronology.* Among his papers apparently in the hand of the period 1710–15 are two drafts of "The Introduction. Of the times before the Assyrian Empire."

Idolatry had its rise [Newton began] from worshipping ye founders of Cities kingdoms & Empires, & began in Chaldea a little before the days of Abraham, most probably by ye worship of Nimrod founder of several great cities. Till Abrahams days the worship of the true God propagated down from Noah to his posterity continued in Canaan as is manifest by the instance of Melchizedeck but in a little time the Canaanites began to imitate the Chaldees in worshipping the founders of their dominions, calling them Baalim & Melchom & Asteroth Lords & Kings & Queens, & sacrificing to them upon their gravestones & in their sepulchres & directing their worship to their statues as their representatives, & instituting colleges of Priests with sacred rites to perpetuate their worship.

At the time when Newton wrote this passage he was master of the Mint, with an annual income of about £2,000, a trusted adviser to the government, and the honored president of the Royal Society who was privileged to refer in print to the princess of Wales as "a particular friend." He has not told us why he suppressed the projected introduction and disguised the central theme of his chronological interests, but it is not far-fetched to speculate that he feared to endanger his position by revealing too much. A dramatic slip of the pen about this time portrays his situation. Intending to write "St John," he put down instead "Sr John." In truth, the Sir Johns of Augustan England bulked larger in his life by 1715 than the St. Johns of the primitive church. The man who had once prepared to surrender his fellowship so as not to accept the mark of the beast now cultivated the odor of orthodox sanctity by serving as a trustee of the Tabernacle on Golden Square and as a member of the Committee to Build Fifty New Churches in London. He was not likely to publish a tract that would place him beyond the pale.

Isaac Newton historian was Isaac Newton heretic engaged in one of his characteristic lifelong activities, the concealment of his heterodox views. In this he was eminently successful. In the continuing comments on the *Chronology* in the eighteenth century, only Arthur Young appears to have perceived Newton's drift. There was indeed only one slip in the whole performance. By concealing his true purpose so effectively, Newton produced a book with no evident point and no evident form. A work of colossal tedium, it excited for a brief time the interest and opposition of the handful able to get excited over the date of the Argonauts before it sank into oblivion. It is read today only by the tiniest remnant who for their sins must pass through its purgatory.

After Newton's death, William Whiston asserted that Newton and Samuel Clarke had given up the good fight for the restoration of primitive Christianity because Newton's interpretation of the prophecies led them to expect a long age of corruption before it would take place. Whiston's assertion pointed correctly to Newton's resumption of his study of the prophecies, which together with his allied interest in ancient chronology became his principal study. During his final decade Newton composed two slightly different expositions of the prophecies, as effectively cleansed of matter likely to give offense as the *Chronology* had been. Conduitt called it "Revelation & Prophecies without *Enthusiasm or superstition* . . . ." Unfortunately, we must add, as we did with the *Chronology*, prophecies without focus or point, but his heirs could publish the manuscripts without concern, as they did, melding the two together without much concern for continuity as the *Observations upon the Prophecies of Daniel, and the Apocalypse of St. John*.

Both the *Chronology* and the *Observations upon the Prophecies* appeared in print after Newton's death. Other theological papers, which were also products of his old age, did not. They provide a perspective by which we can judge the extent to which Newton edited those which he prepared for publication. Among the papers, one of the most significant, which exists in multiple drafts like everything to which Newton attached importance, bore the title "Irenicum." "Irenicum" returned to the theme of the "Origines": "All nations were originally of one religion & this religion consisted in the Precepts of the sons of Noah . . . ." The principal heads of the primitive religion were love of God and love of neighbor. This religion descended to Abraham, Isaac, and Jacob. Moses delivered it to Israel. Pythagoras learned it in his travels and taught it to his disciples. "This religion," Newton concluded, "may be therefore called the Moral Law of all nations." To the two great commandments of love in the primitive religion, the Gospels added the further doctrine that Jesus was the Christ foretold in prophecy. When Jesus was asked what was the great commandment of the law, he answered that it was to love God, and he added that the second commandment was to love your neighbor. "This was the religion of the sons of Noah established by Moses & Christ & still in force." All this was taught from the beginning in the primitive church. To impose now any article of communion that was not such from the beginning was to preach another gospel. To persecute Christians for not receiving that Gospel was to make war on Christ. The two great commandments, he insisted over and over, "always have & always will be the duty of all nations & The coming of Jesus Christ has made no alteration in them."

Somehow Newton blended this attenuated vision of Christianity with a living faith in the almighty God which suffused his life.

We must beleive that there is one God or supreme Monarch that we may fear & obey him & keep his laws & give him honour & glory. We must beleive that he is the father of whom are all things, & that he loves his people as his children that they may mutually love him & obey him as their father. We must beleive that he is the παντοκρατωρ Lord of all things with an irresistible & boundles power & dominion that we may not hope to escape if we rebell & set up other Gods or transgress the laws of his monarchy, & that we may expect great rewards if we do his will. We must beleive that he is the God of the Jews who created the heaven & earth all things therein as is exprest in the ten commandments that we may thank him for our being, & for all the blessings of this life, & forbear to take his name in vain or worship images or other Gods. We are not forbidden to give the name of Gods to Angels & Kings, but we are forbidden to have them as Gods in our worship. For tho there be that are called God whether in heaven or in earth (as there are Gods many & Lords many, yet to us there is but one God the father of whom are all things & we in him & one Lord Jesus Christ by whom are all things & we by him: that is, but one God & one Lord in oͬ worship.

The concept of pantocrator caught Newton's imagination and held it. The word appeared repeatedly throughout the theological papers from his final years. Autocrat over all that is, He dictated the form of the natural world and the course of human history. Newton did not meet Him in the intimacies of watchful providence, a point related to his Arianism. Rather he found Him in the awful majesty of universal immutable laws – an austere God, one perhaps whom only a philosopher could worship.

Whereas Newton published statements of his belief in God, he not only kept the unorthodox aspects of his religion to himself, but he also exercised some care in London to mask his heterodoxy behind a façade of public conformity. He continued to act as a trustee of Archbishop Tenison's chapel on Golden Square until 1722. When Parliament passed an act in 1711 to finance the construction of fifty new churches in the expanding suburbs of London, Newton became one of the commissioners appointed to implement Parliament's will, and he sat on the commission until at least 1720. Likewise he accepted membership on the new commission to supervise the completion of St. Paul's cathedral and attended meetings of it in the period 1715–21. Only a small number knew the true reality. Well concealed, his heterodoxy slid into virtual oblivion, not to be uncovered until the twentieth century or to be fully revealed until the Yahuda papers became available quite recently.

His conclusions functioned only vicariously in the religious ferment of the eighteenth century. When Joseph Hallet, alarmed by the spread of Arianism, published in 1735 *An Address to Conforming Arians* to convince them of their hypocrisy and to lead them to repent, he named two men as the source of the infection, William Whiston and Samuel Clarke. Both were Newton's disciples and known as such. But Newton's extended quest, barely hinted at in his published works, had to enter the stream of religious controversy through disciples more daring than he. He carefully laundered what he himself prepared for publication. The rest he locked away. It is wholly unlikely that his views, formulated a generation before similar ones became widespread, had a significant causal role in the religious history of the Enlightenment.

ॐ

In 1717, while Newton brought out a new edition of his *Opticks,* and while the priority dispute temporarily slumbered following Leibniz's death, a young man who would figure prominently in his declining years entered his life. On 26 August 1717, John Conduitt married Newton's niece Catherine Barton. Conduitt had functioned as commissary to the British forces on the new base at Gibraltar from April 1713 until at least July 1715, more probably until early 1717. When he was the husband of Catherine Barton, Conduitt was a wealthy man. Clearly he had been born to prosperity, and he may have inherited his wealth. Positions such as commissary had traditionally yielded immense profits, however, and it is not implausible to speculate that the years at Gibraltar enhanced Conduitt's means at the least. He did more than accumulate money while he was there. He also identified the site of the Roman city Carteia. On 20 June 1717, with Newton in the chair, Conduitt read a paper on Carteia to the Royal Society. It so happened that Newton was also interested in Carteia, a city built by the Tyrians, as he believed, during humankind's expansion through the Mediterranean basin in the first millennium B.C. We do not know the course of events following Conduitt's appearance at the Royal Society. Newton was working on his *Chronology* in that period, however, and he might well have spoken to the author of a paper that fit in with his current studies. Three months later Conduitt did for Catherine Barton what Halifax had not; he gave her a new name and married respectability. Conduitt was then twenty-nine years old, and Mrs. Catherine Barton was thirty-eight. The evidence indicates that she was still beautiful and charming. Her uncle may nevertheless have possessed at least as much attraction as the bride, for Conduitt worshiped him unabashedly as a hero. Though Conduitt's own capabilities may have been limited, he recognized that he stood in the presence of one of the geniuses of all ages, and he vowed to respond adequately to the opportunity. He wrote down accounts of their conversations. He collected anecdotes about Newton. When he died twenty years after his wedding, Conduitt arranged to have his memorial plaque begin with a statement, not about himself, not about his wife, not about his parents, but about Isaac Newton, to whom he was related and near whose remains he had contrived to have his own placed.

Early in 1718, Newton made the acquaintance of William Stukeley, who was practicing medicine in London and joined the Royal Society at that time. Stukeley was regular in his attendance, and he derived from Lincolnshire as well. Newton became friendly with him. When he later moved to Grantham, Stukeley, like Conduitt, made it his business to collect information about Newton. From the two of them comes much of our knowledge of Newton's characteristics in his final years. Newton's life, Conduitt wrote, was "one continued series of labour, patience, humility, temperance, meekness, humanity, beneficence & piety without any tincture of vice . . . ." Such are the fruits of hero worship. Fortunately, in addition to beatifying Newton as a plaster saint, he also recorded a few details. Newton was of middle stature and, in his later years, plump. He had "a very lively piercing eye" and a gracious aspect. His head of hair, as white as snow, was full with no baldness. Even as an old man he retained the bloom and color of youth and all his teeth except one. In reporting Humphrey Newton's story that he saw Newton laugh only once, Stukeley commented that his own experience was otherwise, though as he filled in the details it appeared less different than he claimed.

According to my own observation, tho' Sir Isaac was of a very serious and compos'd frame of mind, yet I have often seen him laugh, and that upon moderate occasions. . . . He usd a good many sayings, bordering on joke and wit. In company he behavd very agreably; courteous, affable, he was easily made to smile, if not to laugh. . . . He could be very agreable in company, and even sometime talkative.

Percival, the tenant at Woolsthorpe, told Spence that Newton was a man of very few words, "that he would sometimes be silent and thoughtful for above a quarter of an hour together, and look all the while almost as if he was saying his prayers: but that when he did speak, it was always very much to the purpose." The comments remind us more of the Newton of Cambridge days than Conduitt's panegyric does.

Perhaps the most prominent feature in Conduitt's recollection of Newton was studiousness. Newton's age of creativity had ended twenty years before Conduitt first met him. After the move to London, he did nothing but reshuffle ideas and themes from his years in Cambridge. Nevertheless, the pattern of a lifetime remained intact. If he could only reshuffle old ideas, at least he could do that, and a life which found adventure in exploring the seas of thought held true to itself until the end. He gave all of his time not devoted to business and what Conduitt called the civilities of life to study, "& he was hardly ever alone without a pen in his hand & a book before him . . . ." Conduitt noticed that his eyes never grew tired from reading.

He had friends of a sort he had not known in Cambridge. Conduitt informed Fontenelle that George II and his wife (the former Princess Caroline) showed Newton favor and often admitted him to their presence for hours together. The queen liked to hear arguments on questions of philosophy and divinity and sought Newton's company for that purpose. She even claimed to consider his "Abstract" of chronology, which he had written out for her in his own hand, one of her choicest treasures.

By inference we know something else about Newton during his years of decline, something that harmonizes well with his evident pleasure in being a familiar acquaintance of royalty. He was greatly concerned to leave his image behind him. Not just in his old age, but during the whole of his residence in London, he sat constantly to portraits, so that following the Kneller of 1702 (already Kneller's second) no more than four years passed without a new one. During his final decade, portraiture appears to have become almost a mania. After Kneller painted him in 1702 (Plate 3), Jervas did so in 1703, Gandy in 1706, and Thornhill twice in 1709–10 (Plate 4). In 1714 he sat for a miniature by Richter, and that same year Le Marchand sculpted a bust in ivory. Four years later, in 1718, Le Marchand did a second bust (Plate 5) plus a number of reliefs, and Murray a portrait. Kneller executed his third portrait (for the French scientist Varignon) in 1720, and in the three years before his death in 1723 two more (of which only one apparently survives) for Conduitt. Vanderbank did two portraits in 1725 and a third in 1726 (Plate 6), and Seeman another in 1726. There is testimony to a portrait by Dahl, probably from Newton's final years. Two other portraits of him in old age by unknown artists exist, one in the National Portrait Gallery and one in the possession of W. Heffer and Sons. One or both may be copies; one of them may be the Dahl. Many or perhaps most of these were commissioned by other men, but they could have been carried out only with Newton's cooperation. By any reckoning it is a considerable record; *obsession* does not seem too strong a word.

Another characteristic does not reduce readily to his concern with his image. Charity supplied a constant background to Newton's final years.

He dispensed much of it to various branches of his family, for he was by far the most prosperous man in the clan, the other members of which looked to him. In the early eighteenth century their troubles outnumbered their joys, and they brought the troubles to rich Sir Isaac. When Thomas Pilkington, the husband of his half-sister, died and left Mary Smith Pilkington a widow like her younger sister Hannah, Newton came to her support, and later he was making regular quarterly payments of £9 to sustain her daughter Mary. He stood surety for a loan of £20 to his sister's son, Thomas Pilkington. And there were many others. The onslaughts of fortune brought a constant stream of destitute relations to beg at his door. Nor did he confine his alms to his family. Among his papers are a large number of letters begging assistance. Their very number implies that he was known in circles that cared as a charitable man. The letters contain internal evidence that he answered many appeals. Together with the abiding friendships of his London years, Newton's charity works to soften the image left by the quarrels with Flamsteed and Leibniz. The quarrels were real. So was the charity to unfortunates, as though he hoped to compensate for his own shortcomings. The extensive charity should not be taken to mean that Newton was unconcerned with his own material condition. When he died, he left a considerable estate, and we must assume that he watched over its accumulation carefully during all the years in London.

During his final years, Newton liked to reminisce over the various topics that had formed the substance of his life. At least three people heard the story about the apple and the law of gravity independently. Stukeley engaged him at times on chronology and the prophecies, although Newton never let him glance into the depths of his theological reflections. Conduitt occasionally heard some of his wider-ranging speculations. On 7 March 1725 they had a long conversation, which Conduitt recorded in a memorandum, about circulations in the cosmos. Newton told him his belief that there was a sort of revolution of the heavenly bodies. Light and vapors from the sun gather together to make secondary bodies like the moon, which continue to grow as they gather more matter and become primary planets and ultimately comets, which in turn fall into the sun to replenish its matter. He thought that the great comet of 1680, after five or six more orbits, would fall into the sun, increasing its heat so much that life on earth would cease. Humankind was of recent date, he continued, and there were marks of ruin on the earth which suggested earlier cataclysms like the one he predicted. Conduitt asked how the earth could have been repeopled if life had been destroyed. It required a creator, Newton answered. Why did he not publish his conjectures as Kepler had done? "I do not deal in conjectures," he replied. He picked up the *Principia* and showed Conduitt hints of his belief that he had put in the discussion of comets. Why did he not own it outright? He laughed and said that he had published enough for people to know his meaning.

Not long before his death, Newton looked back over his life and summarized it for some unnamed companion. It is a magnificent reflection which catches the essence of a life devoted to the pursuit of Truth.

I don't know what I may seem to the world, but, as to myself, I seem to have been only like a boy playing on the sea shore, and diverting myself in now and then finding a smoother pebble or a prettier shell than ordinary, whilst the great ocean of truth lay all undiscovered before me.

Signs of senescence though never senility began to appear toward the end. Conduitt, who was usually careful not to reveal any hint of decay in Newton, mentioned in his memorandum of 7 March 1725 that Newton's head was clearer and his memory stronger that day than it had been for some time. Pemberton also noted that Newton's memory was much decayed. Conduitt stated that after 1725 Newton hardly ever went to the Mint and that he himself took over the duties.

In the last five years, Newton's health began visibly to deteriorate. His basic problem, perhaps a result of a serious illness in 1723, was a weakness of the sphincter. From this time on, Newton suffered from incontinence of urine. Because motion excited the affliction, he gave up his carriage and took to a chair. He quit dining abroad and seldom entertained at home. He stopped eating flesh in any quantity and lived chiefly on broth, vegetables, and soup. In January 1725, Newton suffered a violent cough and inflammation of the lungs. An attack of gout further compounded his problems. After 7 January, he did not occupy the chair of the Royal Society again until 22 April, and from that time until his death he missed more meetings than he attended. After much ado, the Conduitts persuaded him to take a house in Kensington. The air proved to be good for him. Conduitt noted that he was "visibly better" than he had been for some years, an implicit admission of greater decay than Conduitt ever stated openly. His spirit never deserted him. Conduitt tried to get him not to walk to church, to which he replied, "use legs & have legs." Conduitt also remarked that he continued to study and write until the very end.

A few days before his death, Zachary Pearce, rector of Newton's home parish, St. Martins-in-the-Fields, visited him.

I found him writing over his *Chronology of Ancient Kingdoms,* without the help of spectacles, at the greatest distance of the room from the windows, and with a parcel of books on the table, casting a shade upon the paper. Seeing this, on my entering the room, I said to him, "Sir, you seem to be writing in a place where you cannot so well see." His answer was, "A little light serves me." He then told me that he was preparing his Chronology for the press, and that he had written the greatest part of it over again for that purpose. He read to me two or three sheets of what he had written, (about the middle, I think, of the work) on occasion of some points in Chronology, which had been mentioned in our conversation. I believe, that he continued reading to me, and talking about what he had read, for near an hour, before the dinner was brought up.

After the summer break in 1726, Newton attended only four meetings of the Royal Society and only one meeting of the council. He presided for the last time on 2 March 1727. Exhilarated by the meeting, he stayed on in London that night, and the next day Conduitt thought he had not seen him better in many years. But the strain both of the meeting and of the visits he received on the morrow brought his violent cough back, and he returned to Kensington on 4 March. Conduitt sent for Mead and Cheselden, two prominent physicians who tended Newton. They diagnosed the distemper as a stone in the bladder and offered no hope of recovery. Newton was in great pain. In his anguish sweat ran from his face. The spectacle of Christian dying was more than Stukeley's poetic gift could resist.

It [the pain] rose to such a height that the bed under him, and the very room, shook with his agonys, to the wonder of those that were present. Such a struggle had his great soul to quit its earthly tabernacle! All this he bore with a most exemplary and remarkable patience, truly philosophical, truly Christian . . . .

Conduitt felt obliged to note, not without embarrassment, what was surely Newton's most significant act as he lay dying, though Stukeley would have found it difficult to reconcile with his account. Newton refused to receive the sacrament of the church. He must have been plan-ning the gesture for some time, the personal declaration of belief he had not dared to utter in public for more than fifty years. Even in death it was compromised. He had, after all, spent his final fifteen years purging objectionable opinions from the theological works he left behind for publication. Similarly, he made his gesture to the limited audience of Catherine and John Conduitt, who did not care to jeopardize his memory by making it known. As far as one can tell, Stukeley never heard a whisper about it.

On 15 March Newton was somewhat better, and the watchers began to hope. He declined again immediately, however, was insensible on Sunday, 19 March, and died the following morning about one o'clock without further pain.

He may not have given the Royal Society much leadership during his final years, but they were aware of what they had lost.

March 23^d. 1726.
The Chair being Vacant by the Death of Sir Isaac Newton there was no Meeting this Day.

੨ॖ

Death has its unavoidable practical dimension. The liquid estate, made up primarily of stock and annuities in the Bank of England and the South Sea Company, amounted to £32,000, not a princely sum but a very substantial one. Eight nieces and nephews of half blood – three Pilkingtons, three Smiths, and two Bartons (including Catherine Barton Conduitt, of course) – shared it. Among them, only Catherine Conduitt, seconded by her husband, appreciated Newton's achievement. To the others he was only rich Uncle Isaac, and they determined to realize as much as they could from their good fortune. Inevitably there was some bickering, tempered in the end by a hard reality. Before the courts would allow the estate to be distributed, someone would have to assume the obligation of the master of the Mint's accounts, assuring the crown against loss. Among the heirs, only Conduitt was in a position to shoulder such a burden. Hence a bargain was struck, according to which all of the manuscripts deemed worthy of publication would be printed and sold "to the best advantage." In the meantime, Catherine Conduitt would keep them. Conduitt agreed to pass the accounts. In return, he would have possession of all the papers not found worthy of publication. They agreed also, no doubt at Conduitt's prodding, to set aside £500 from the estate for a monument. They spent £87 on 101 funeral rings, sent £20 to the poor in Colsterworth parish, and gave every servant a year's wages.

Three of the manuscripts did see print, the *Chronology* and *Observations upon the Prophecies*, as I have mentioned, and Newton's original draft of the final book of the *Principia*, which was published under the title *De mundi systemate* in 1728. Through the agreement Conduitt obtained effective ownership of the papers. Conduitt was adroit enough also to secure much the richest inheritance, the mastership of the Mint, for himself.

Stukeley asserted that each of the eight nieces and nephews inherited about £3,500 (which understated it by nearly £500), "but all soon found a period." Little is known about seven of them, but Stukeley's summary judgment does not suffice for Catherine Conduitt. She died in 1739, two years after John. Their daughter Catherine married the Hon. John Wallop, Viscount Lymington, in 1740, and their son became the second Earl of Portsmouth. Through Catherine Conduitt, the daughter, Newton's papers, preserved by her father's farsightedness, passed into the possession of the Portsmouth family, and most of them eventually into the Cambridge University Library.

Newton's death scarcely passed unnoticed. The various gazettes carried announcements of it. *The Political State of Great Britain* for March devoted three pages to an encomium which adequately summarized the position Newton held in English learning by proclaiming him "the greatest of Philosophers, and the Glory of the British Nation." James Thomson's "Poem Sacred to the Memory of Sir Isaac Newton" reached a fifth edition before the year was out. The nation that had honored him with knighthood honored him more lavishly in death. On 28 March he lay in state in the Jerusalem Chamber in Westminster Abbey, whence he was interred in a prominent position in the nave. Conduitt informed Fontenelle that the dean and chapter had often refused that place to the greatest noblemen. According to the proprieties of the day, a Knight of the Bath named Sir Michael Newton performed as chief mourner followed by some other relations, as they were called, and eminent persons acquainted with Newton. The Lord Chancellor, the dukes of Montrose and Roxburgh, and the earls of Pembroke, Sussex and Macclesfield, all of them members of the Royal Society, supported the pall; the Bishop of Rochester attended by the prebends and choir performed the office. The monument specified by the heirs was finally erected in 1731, a baroque monstrosity with cherubs holding emblems of Newton's discoveries, Newton himself in a reclining posture, and a female figure representing Astronomy, the Queen of the Sciences, sitting and weeping on a globe that surmounts the whole. Twentieth-century taste runs along simpler lines, and the monument is now roped off in the Abbey so that one can scarcely even see it. A similar tone marked the inscription, which concluded with the exhortation "Let Mortals rejoice That there has existed such and so great an Ornament to the Human Race." In this case, baroque extravagance struck the proper note. Faults Newton had in abundance. Nevertheless, only hyperbole can hope to express the reality of the man who returned to dust in the early spring of 1727.

Bibliographical Essay

L IKE THE BOOK AS A WHOLE, the bibliography is radically condensed and confined to titles in English. Anyone desiring a more
extensive guide to the literature on Newton can find it in *Never at Rest*.

Newton has been the subject of many biographers. The reigning biog-
raphy for many years (a book that will always be worthy of attention) was
David Brewster, *Memoirs of the Life, Writings, and Discoveries of Sir Isaac
Newton*, 2 vols. (Edinburgh: Thomas Constable, 1855). Augustus De
Morgan published a number of pieces criticizing Brewster's excessive
hero worship. Three of them were collected together and edited by Philip
E.B. Jourdain in *Essays on the Life and Work of Newton* (Chicago: Open
Court, 1914). Louis Trenchard More, *Isaac Newton: A Biography* (New
York: Charles Scribner's Sons, 1934), failed to add anything new and also
failed in its attempt to supplant Brewster as the leading biography. J.W.N.
Sullivan, *Isaac Newton, 1642–1727* (London: Macmillan, 1938), a shorter
work, contains general interpretive ideas that continue to merit attention.
Among the large number of popular biographies, unquestionably the best
are E. N. da C. Andrade, *Isaac Newton* (London: Parrish, 1950; and a
slightly different version, *Sir Isaac Newton* [London: Collins, 1954]), and
the longer, more detailed Gale E. Christianson, *In the Presence of the
Creator. Isaac Newton and His Times* (New York: Free Press, 1984). No
survey of biographies of Newton can omit Frank E. Manuel, *Portrait of
Isaac Newton* (Cambridge, Mass.: Harvard University Press, 1968). Because it does not attempt to deal with Newton's scientific career, it is not a
biography in the full sense of the word. On every other aspect of Newton's
life it offers insights no student of Newton can afford to ignore. It also
offers a Freudian analysis of the roots of Newton's character which may or

may not be true but can be separated from the portrait of Newton, on which it is empirically based.

There are many studies of Newton's work and thought in general. E. A. Burtt, *The Metaphysical Foundations of Modern Physical Science* (New York: Harcourt Brace, 1925), concludes with a long discussion of Newton which, along with the book as a whole, deservedly continues to exercise influence. I. Bernard Cohen, the dean of practicing Newtonian scholars, effectively began the Newtonian aspect of his career with his important *Franklin and Newton, an Inquiry into Speculative Newtonian Experimental Science and Franklin's Work in Electricity as an Example Thereof* (Philadelphia: American Philosophical Society, 1956). As the title indicates, the primary focus of Cohen's work falls beyond Newton; nevertheless, it begins with a long discussion of him as the basis of the rest. Another of Cohen's contributions to the understanding of Newton will appear later in this essay. Here let me add *The Birth of a New Physics* (Garden City, N.Y.: Anchor, 1960). Alexandre Koyré, who did so much to shape the modern discipline of the history of science, turned his attention to Newton in his late years. See especially his *Newtonian Studies* (Cambridge, Mass.: Harvard University Press, 1965). One of the *Newtonian Studies* deserves separate mention: "The Significance of the Newtonian Synthesis." See also the parts devoted to Newton in Koyré's *From the Closed World to the Infinite Universe* (Baltimore: Johns Hopkins University Press, 1965). Henry Guerlac also contributed to our understanding of Newton during the later part of his career. See his *Newton et Epicure, Conférence donnée au Palais de la Découverte le 2 Mars 1963* (Paris: Conférences du Palais de la Découverte, 1963), and "Newton's Optical Aether," *Notes and Records of the Royal Society, 22* (1967), 45–57. In addition to A. R. and Marie Boas Hall's "Newton's Theory of Matter," *Isis, 51* (1960), 131–44, see their introductions in *Unpublished Scientific Papers of Isaac Newton* (Cambridge: Cambridge University Press, 1962). B.J.T. Dobbs, *The Janus Faces of Genius* (New York: Cambridge University Press, 1992), appearing just as I compose this bibliography, is certain to be recognized as one of the most important interpretations of Newton.

To the extent that they relied on the published works of Newton, all of the studies done before 1945 have been somewhat superseded by more recent ones (such as most of those in the foregoing paragraphs), which are based primarily on his manuscripts. One of the earliest of these, and one

that broke decisively with the established pattern of apotheosizing Newton, was Lord Keynes's essay drawn from his own collection of papers:
"Newton, the Man," in the Royal Society's volume *Newton Tercentenary
Celebrations* (Cambridge: Cambridge University Press, 1947), pp. 27–34.
J. E. McGuire has utilized the manuscript sources extensively in a
number of influential articles on Newton; see especially the paper written
with P. M. Rattansi, "Newton and the 'Pipes of Pan'," *Notes and Records of
the Royal Society*, *21* (1966), 108–43. Analogous to McGuire in its use of
Newton's manuscripts as a means of examining philosophical questions in
his science is Ernan McMullin, *Matter and Activity in Newton* (Notre
Dame, Ind.: University of Notre Dame Press, 1977). Another important
article is David Kubrin, "Newton and the Cyclical Cosmos: Providence
and the Mechanical Philosophy," *Journal of the History of Ideas*, *28* (1967),
325–46.

   D. T. Whiteside has established himself as the unquestioned authority
on Newton's mathematics. He should be consulted first of all in the
introductory essays and editorial apparatus of his great edition of the
Mathematical Papers, and in the introduction to a reprint edition of the six
mathematical papers published during Newton's life or shortly thereafter:
The Mathematical Works of Isaac Newton, 2 vols. (New York: Johnson Reprint Corp., 1964). Beyond these, he has published a number of important works and articles, among which I will cite "Isaac Newton: Birth of a
Mathematician," *Notes and Records of the Royal Society*, *19* (1964), 53–62.
Carl B. Boyer, *The Concepts of the Calculus: A Critical and Historical Discus-
sion of the Derivative and the Integral* (New York: Columbia University
Press, 1939), also contains an excellent discussion of Newton. The best
narrative of the priority dispute with Leibniz is A. R. Hall, *Philosophers at
War: The Quarrel between Newton and Leibniz* (Cambridge: Cambridge
University Press, 1980).

   Among studies of Newton's optics, A. R. Hall effectively inaugurated
the recent heavy exploitation of the Newtonian manuscripts with his article "Sir Isaac Newton's Note-book, 1661–65," *Cambridge Historical Journal*, *9* (1948), 239–50, which he followed with "Further Optical Experiments of Isaac Newton," *Annals of Science*, *11* (1955), 27–43. There is an
extensive passage on Newton in A. I. Sabra, *Theories of Light from Descartes
to Newton* (London: Oldbourne, 1967). Thomas S. Kuhn's introductory
essay to "Newton's Optical Papers," in I. B. Cohen's volume of *Papers &*

Letters, is excellent. Alan E. Shapiro, who is editing Newton's *Optical Papers,* has become the recognized authority on this subject. Among his numerous articles see "The Evolving Structure of Newton's Theory of White Light and Color: 1670–1704," *Isis, 71* (1980), 211–35.

The *Principia* and related questions in mechanics have been the subject of historical study for longer than the optics. I. Bernard Cohen has made the *Principia* his special province in works too extensive to be cited here in entirety, but see especially his *Introduction to Newton's 'Principia'* (Cambridge: Cambridge University Press, 1971), an invaluable history of the book itself. John Herivel is the leading student of the early development of Newton's mechanics; his articles on this subject are effectively embodied in the various essays included in *The Background to Newton's 'Principia'* (Oxford: Oxford University Press, 1965), which also publishes all of the documents on Newton's mechanics before the *Principia.*

Newton's chemistry (and alchemy) have been much less studied than other aspects of his science. B.J.T. Dobbs, *The Foundations of Newton's Alchemy: The Hunting of the Greene Lyon* (Cambridge: Cambridge University Press, 1975), supplants all earlier work and, for the first time, offers real guidance into the comprehension of the extensive alchemical papers he left behind.

Frank E. Manuel, *The Religion of Isaac Newton* (Oxford: Oxford University Press, 1974), is the only work on Newton's religion of which I am aware that draws upon the Yahuda manuscripts, which have only recently become available to scholars. Manuel concerns himself entirely, however, in extending the Freudian themes of his *Portrait,* so that the book does not give any insight into the theological content of the papers. Manuel's *Isaac Newton, Historian* (Cambridge, Mass.: Harvard University Press, 1963), is the only significant study of Newton's chronology, a topic intimately allied with his theology. Margaret Jacob, *The Newtonians and the English Revolution, 1689–1720* (Ithaca, N.Y.: Cornell University Press, 1976), which sums up a number of earlier articles, examines the interrelation of Newtonian natural philosophy, the practical theology of the Latitudinarians, and the political situation in England at the time of the Glorious Revolution.

Index

319